PELICAN BOOKS

INVITATION TO SOCIOLOGY

Peter L. Berger is at present Professor of Sociology at Rutgers University, New York. Until 1970 he was Professor of Sociology in the Graduate Faculty of the New School for Social Research and editor of its quarterly, *Social Research*. His publications in the field of religion include *The Noise of Solemn Assemblies* and *The Social Reality of Religion*. His best-known book on sociology is *Invitation to Sociology – A Humanistic Perspective* and he is co-author, with Thomas Luckmann, of *The Social Construction of Reality – A Treatise in the Sociology of Knowledge* (Allen Lane The Penguin Press, 1969).

PETER L. BERGER

INVITATION TO SOCIOLOGY

A HUMANISTIC PERSPECTIVE

PENGUIN BOOKS

Penguin Books Ltd, Harmondsworth, Middlesex, England
Penguin Books Australia Ltd. Ringwood Victoria, Australia

—

First published in the U.S.A. by Doubleday & Co., 1963
Published in Pelican Books 1966
Reprinted 1967, 1968, 1969, 1970, 1971 (twice), 1972, 1973

—

Copyright © Peter L. Berger 1963

—

Made and printed in Great Britain by
Hunt Barnard Printing Ltd, Aylesbury
Set in Linotype Times

CONTENTS

PREFACE

THIS book is intended to be read, not studied. It is not a textbook or an attempt at theoretical system-building. It is an invitation to an intellectual world that I consider to be profoundly exciting and significant. In issuing such an invitation it is necessary to delineate the world to which the reader is being invited, but it will be clear that the latter will have to go beyond this book if he decides to take the invitation seriously.

In other words, the book is addressed to those who, for one reason or another, have come to wonder or to ask questions about sociology. Among these, I have supposed, will be students who may be toying with the idea of taking up sociology in a serious way, as well as more mature members of that somewhat mythological entity called the 'educated public'. I have also supposed that some sociologists may be attracted to the book, although it will tell them few things that they don't know already, since all of us derive a certain narcissistic satisfaction in looking at a picture that includes ourselves. Since the book is addressed to a fairly wide audience, I have avoided as much as possible the technical dialect for which sociologists have earned a dubious notoriety. At the same time, I have avoided talking down to the reader – mainly because I regard this as a repulsive stance in itself, but also because I don't particularly want to invite to this game people, including students, whom one feels constrained to talk down to. I shall admit frankly that, among the academic diversions available today, I consider sociology as a sort of 'royal game' – one doesn't invite to a chess tournament those who are incapable of playing dominoes.

It is inevitable that such an undertaking will reveal the writer's prejudices concerning his own field. This too must be admitted frankly from the beginning. If other sociologists should read this book, especially in America, some will unavoidably be irritated by its orientation, disapprove of some of its lines of argument and feel that things considered by them to be important have been left out. All I can say is that I have tried to be faithful to a central tradition that goes back to the classics in the field and that I believe strongly in the continuing validity of this tradition.

My special prepossession in the field has been the sociology of religion. This will perhaps be evident from the illustrations that I use because they come most readily to my mind. Beyond that, however, I have tried to avoid an emphasis on my own speciality. I have wanted to invite the reader to a rather large country, not to the particular hamlet in which I happen to live.

In writing this book I was faced with the choice of inserting thousands of footnotes or none at all. I decided on the latter course, feeling that little would be gained by giving the book the appearance of a Teutonic treatise. In the text, names are given where ideas are not part of the broad consensus in the field. These names are taken up again in the bibliographical comments at the end of the book, where the reader will also find some suggestions concerning further readings.

In all my thinking on my chosen field I owe an immense debt of gratitude to my teacher Carl Mayer. If he should read this book, I suspect that there will be passages that will make him raise an eyebrow. I still hope that he would not regard the conception of sociology here presented as too much of a travesty on the one he has been conveying to his students. In one of the subsequent chapters I take the position that all world views are the result of conspiracies. The same can be said of views concerning a scholarly discipline. In conclusion, then, I would like to thank three individuals who have been fellow conspirators through many conversations and arguments – Brigitte Berger, Hansfried Kellner and Thomas Luckmann. They will find the results of these occasions in more than one place in the following pages.

Hartford, Connecticut

P.L.B.

1
SOCIOLOGY AS AN INDIVIDUAL PASTIME

THERE are very few jokes about sociologists. This is frustrating for the sociologists, especially if they compare themselves with their more favoured second cousins, the psychologists, who have pretty much taken over that sector of American humour that used to be occupied by clergymen. A psychologist, introduced as such at a party, at once finds himself the object of considerable attention and uncomfortable mirth. A sociologist in the same circumstance is likely to meet with no more of a reaction than if he had been announced as an insurance salesman. He will have to win his attention the hard way, just like everyone else. This is annoying and unfair, but it may also be instructive. The dearth of jokes about sociologists indicates, of course, that they are not as much part of the popular imagination as psychologists have become. But it probably also indicates that there is a certain ambiguity in the images that people do have of them. It may thus be a good starting point for our considerations to take a closer look at some of these images.

If one asks undergraduates why they are taking sociology as a main subject, one often gets the reply, 'because I like to work with people'. If one then goes on to ask such students about their occupational future, as they envisage it, one often hears that they intend to go into social work. Of this more in a moment. Other answers are more vague and general, but all indicate that the student in question would rather deal with people than with things. Occupations mentioned in this connexion include personnel work, human relations in industry, public relations, advertising, community planning or religious work of the unordained variety. The common assumption is that in all these lines of endeavour one might 'do something for people', 'help people', 'do work that is useful for the community'. The image of the sociologist involved here could be described as a secularized version of the liberal Protestant ministry, with the Y M C A secretary perhaps furnishing the connecting link between sacred and profane benevolence. Sociology is seen as an up-to-date variation on the classic American theme of 'uplift'. The sociologist is understood as one professionally concerned with edifying activities on behalf of individuals and of the community at large.

One of these days a great American novel will have to be written on the savage disappointment this sort of motivation is bound to suffer in most of the occupations just mentioned. There is moving pathos in the fate of these likers of people who go into personnel work and come up for the first time against the human realities of a strike that they must fight on one side of the savagely drawn battle lines, or who go into public relations and discover just what it is that they are expected to put over in what experts in the field have called 'the engineering of consent', or who go into community agencies to begin a brutal education in the politics of real estate speculation. But our concern here is not with the despoiling of innocence. It is rather with a particular image of the sociologist, an image that is inaccurate and misleading.

It is, of course, true that some Boy Scout types have become sociologists. It is also true that a benevolent interest in people could be the biographical starting point for sociological studies. But it is important to point out that a malevolent and misanthropic outlook could serve just as well. Sociological insights are valuable to anyone concerned with action in society. But this action need not be particularly humanitarian. Some American sociologists today are employed by governmental agencies seeking to plan more

livable communities for the nation. Other American sociologists are employed by governmental agencies concerned
with wiping communities of hostile nations off the map, if
and when the necessity should arise. Whatever the moral
implications of these respective activities may be, there is
no reason why interesting sociological studies could not be
carried on in both. Similarly, criminology, as a special field
within sociology, has uncovered valuable information about
processes of crime in modern society. This information is
equally valuable for those seeking to fight crime as it would
be for those interested in promoting it. The fact that more
criminologists have been employed by the police than by
gangsters can be ascribed to the ethical bias of the criminologists themselves, the public relations of the police and
perhaps the lack of scientific sophistication of the gangsters.
It has nothing to do with the character of the information
itself. In sum, 'working with people' can mean getting them
out of slums or getting them into jail, selling them propaganda or robbing them of their money (be it legally or
illegally), making them produce better automobiles or
making them better bomber pilots. As an image of the
sociologist, then, the phrase leaves something to be desired,
even though it may serve to describe at least the initial
impulse as a result of which some people turn to the study
of sociology.

Some additional comments are called for in connexion
with a closely related image of the sociologist as a sort of
theoretician for social work. This image is understandable
in view of the development of sociology in America. At
least one of the roots of American sociology is to be found
in the worries of social workers confronted with the massive
problems following in the wake of the industrial revolution
– the rapid growth of cities and of slums within them, mass
immigration, mass movements of people, the disruption of
traditional ways of life and the resulting disorientation of

individuals caught in these processes. Much sociological research has been spurred by this sort of concern. And so it is still quite customary for undergraduate students planning to go into social work to major in sociology.

Actually, American social work has been far more influenced by psychology than by sociology in the development of its 'theory'. Very probably this fact is not unrelated to what was previously said about the relative status of sociology and psychology in the popular imagination. Social workers have had to fight an uphill battle for a long time to be recognized as 'professionals', and to get the prestige, power and (not least) pay that such recognition entails. Looking around for a 'professional' model to emulate, they found that of the psychiatrist to be the most natural. And so contemporary social workers receive their 'clients' in an office, conduct fifty-minute 'clinical interviews' with them, record the interviews in quadruplicate and discuss them with a hierarchy of 'supervisors'. Having adopted the outward paraphernalia of the psychiatrist, they naturally also adopted his ideology. Thus contemporary American social-work 'theory' consists very largely of a somewhat bowdlerized version of psychoanalytic psychology, a sort of poor man's Freudianism that serves to legitimate the social worker's claim to help people in a 'scientific' way. We are not interested here in investigating the 'scientific' validity of this synthetic doctrine. Our point is that not only does it have very little to do with sociology, but it is marked, indeed, by a singular obtuseness with regard to social reality. The identification of sociology with social work in the minds of many people is somewhat a phenomenon of 'cultural lag', dating from the period when as yet pre-'professional' social workers dealt with poverty rather than with libidinal frustration, and did so without the benefit of a dictaphone.

But even if American social work had not jumped on the bandwagon of popular psychologism the image of the sociologist as the social worker's theoretical mentor would be misleading. Social work, whatever its theoretical rationalization, is a certain *practice* in society. Sociology is not a practice, but an *attempt to understand*. Certainly this understanding may have use for the practitioner. For that matter, we would contend that a more profound grasp of sociology would be of great use to the social worker and that such grasp would obviate the necessity of his descending into the mythological depths of the 'subconscious' to explain matters that are typically quite conscious, much more simple and, indeed, *social* in nature. But there is nothing inherent in the sociological enterprise of trying to understand society that necessarily leads to this practice, or to any other. Sociological understanding can be recommended to social workers, but also to salesmen, nurses, evangelists and politicians – in fact, to anyone whose goals involve the manipulation of men, for whatever purpose and with whatever moral justification.

This conception of the sociological enterprise is implied in the classic statement by Max Weber, one of the most important figures in the development of the field, to the effect that sociology is 'value-free'. Since it will be necessary to return to this a number of times later, it may be well to explicate it a little further at this point. Certainly the statement does *not* mean that the sociologist has or should have no values. In any case, it is just about impossible for a human being to exist without any values at all, though, of course, there can be tremendous variation in the values one may hold. The sociologist will normally have many values as a citizen, a private person, a member of a religious group or as an adherent of some other association of people. But within the limits of his activities as a sociologist there is one fundamental value only – that of scientific integrity. Even there, of course, the sociologist, being human, will have to reckon with his convictions, emotions and prejudices. But it is part of his intellectual training that he tries to understand and control these as *bias* that ought to be eliminated, as far as possible, from his work. It goes without saying that this is not always easy to do, but it is not impossible. The sociologist tries to see what is there. He may have hopes or fears concerning what he may find. But he will try to see regardless of his hopes or fears. It is thus an act of pure perception, as pure as humanly limited means allow, toward which sociology strives.

An analogy may serve to clarify this a little more. In any political or military conflict it is of advantage to capture the information used by the intelligence organs of the opposing side. But this is so only because good intelligence consists of information free of bias. If a spy does his reporting in terms of the ideology and ambitions of his superiors, his reports are useless not only to the enemy, if the latter should capture them, but also to the spy's own side. It has been claimed that one of the weaknesses of the espionage apparatus of totalitarian states is that spies report not what they find but what their superiors want to hear. This, quite evidently, is bad espionage. The good spy reports what is there. Others decide what should be done as a result of his information. The sociologist is a spy in very much the same way. His job is to report as accurately as he can about a certain social terrain. Others, or he himself in a role other than that of sociologist, will have to decide what moves ought to be made in that terrain. We would stress strongly that saying this does *not* imply that the sociologist has no responsibility to ask about the goals of his employers or the use to which they will put his work. But this asking is not sociological asking. It is asking the same questions that any man ought to ask himself about his actions in society. Again, in the same way, biological knowledge can be employed to heal or to kill. This does not mean that the biologist is free of responsibility as to which use he serves. But when he asks himself about this responsibility, he is not asking a biological question.

Another image of the sociologist, related to the two already discussed, is that of social reformer. Again, this image has historical roots, not only in America but also in Europe. Auguste Comte, the early nineteenth-century French philosopher who invented the name of the discipline, thought of sociology as the doctrine of progress, a secularized successor to theology as the mistress of the sciences. The sociologist in this view plays the role of arbiter of all branches of knowledge for the welfare of men. This notion, even when stripped of its more fantastic pretensions, died especially hard in the development of French sociology. But it had its repercussions in America too, as when, in the early days of American sociology, some transatlantic disciples of Comte seriously suggested in a memorandum to the president of Brown University that all the departments of the latter should be reorganized under the department of sociology. Very few sociologists today, and probably none in this country, would think of their role in this way. But something of this conception survives when sociologists are expected to come up with blueprints for reform on any number of social issues.

It is gratifying from certain value positions (including some of this writer's) that sociological insights have served in a number of instances to improve the lot of groups of human beings by uncovering morally shocking conditions or by clearing away collective illusions or by showing that socially desired results could be obtained in more humane fashion. One might point, for example, to some applications of sociological knowledge in the penological practice of Western countries. Or one might cite the use made of sociological studies in the Supreme Court decision of 1954 on racial segregation in the public schools. Or one could look at the applications of other sociological studies to the humane planning of urban redevelopment. Certainly the sociologist who is morally and politically sensitive will derive gratification from such instances. But, once more, it will be well to keep in mind that what is at issue here is not sociological understanding as such but certain applications of this understanding. It is not difficult to see how the same understanding could be applied with opposite intentions. Thus the sociological understanding of the dynamics of racial prejudice can be applied effectively by those promoting intragroup hatred as well as by those wanting to spread tolerance. And the sociological understanding of the nature of human solidarity can be employed in the service of both totalitarian and democratic régimes. It is sobering to realize that the same processes that generate consensus can be manipulated by a social group worker in a summer camp in the Adirondacks and by a Communist brainwasher in a prisoner camp in China. One may readily grant that the sociologist can sometimes be called upon to give advice when it comes to changing certain social conditions deemed undesirable. But the image of the sociologist as social reformer suffers from the same confusion as the image of him as social worker.

If these images of the sociologist all have an element of 'cultural lag' about them, we can now turn to some other images that are of more recent date and refer themselves to more recent developments in the discipline. One such image is that of the sociologist as a gatherer of statistics about human behaviour. The sociologist is here seen essentially as an aide-de-camp to an IBM machine. He goes out with a questionnaire, interviews people selected at random, then goes home, enters his tabulations on to innumerable punch cards, which are then fed into a machine. In all of this, of course, he is supported by a large staff and a very large budget. Included in this image is the implication that the results of all this effort are picayune, a pedantic restatement of what everybody knows anyway. As one observer remarked pithily, a sociologist is a fellow who spends $100,000 to find his way to a house of ill repute.

This image of the sociologist has been strengthened in the public mind by the activities of many agencies that might well be called para-sociological, mainly agencies concerned with public opinion and market trends. The pollster has become a well-known figure in American life, importuning people about their views from foreign policy to toilet paper. Since the methods used in the pollster business bear close resemblance to sociological research, the growth of this image of the sociologist is understandable. The Kinsey studies of American sexual behaviour have probably greatly augmented the impact of this image. The fundamental sociological question, whether concerned with premarital petting or with Republican votes or with the incidence of gang knifings, is always presumed to be 'how often?' or 'how many?' Incidentally the very few jokes current about sociologists usually relate to this statistical image (which jokes had better be left to the imagination of the reader).

Now it must be admitted, albeit regretfully, that this image of the sociologist and his trade is not altogether a product of fantasy. Beginning shortly after World War I, American sociology turned rather resolutely away from theory to an intensive preoccupation with narrowly circumscribed empirical studies. In connexion with this turn, sociologists increasingly refined their research techniques. Among these, very naturally, statistical techniques figured prominently. Since about the mid 1940s there has been a revival of interest in sociological theory, and there are good indications that this tendency away from a narrow empiricism is continuing to gather momentum. It remains true, however, that a goodly part of the sociological enterprise in this country continues to consist of little studies of obscure fragments of social life, irrelevant to any broader theoretical concern. One glance at the table of contents of the major sociological journals or at the list of papers read at sociological conventions will confirm this statement.

The political and economic structure of American academic life encourages this pattern, and not only in sociology. Colleges and universities are normally administered by very busy people with little time or inclination to delve into the esoterica produced by their scholarly employees. Yet these administrators are called upon to make decisions concerning the hiring and firing, promotion and tenure of their faculty personnel. What criteria should they use in these decisions? They cannot be expected to read what their professors write, having no time for such activities and, especially in the more technical disciplines, lacking the necessary qualifications to judge the material. The opinions of immediate colleagues of the professors in question are suspect *a priori*, the normal academic institution being a jungle of bitter warfare between faculty factions, none of which can be relied upon for an objective judgment of members of either his own or an opposing group. To ask the views of students would be even more uncertain procedure. Thus the administrators are left with a number of equally unsatisfactory options. They can go on the principle that the institution is one happy family, in which every member advances steadily up the status ladder regardless of merit. This has been tried often enough, but becomes ever more difficult in an age of competition for the favour of the public and the funds of foundations. Another option is to rely on the advice of one clique, chosen on some more or less rational basis. This creates obvious political difficulties for the administrator of a group chronically defensive about its independence. The third option, the one most common today, is to fall back on the criterion of productivity as used in the business world. Since it is very difficult indeed to judge the productivity of a scholar with whose field one is not well acquainted, one must somehow try to find out how acceptable the scholar is to unprejudiced colleagues in his field. It is then assumed that such acceptability can be deduced from the number of books or articles that publishers or editors of professional publications are willing to accept from the man in question. This forces scholars to concentrate on work that can be easily and speedily be converted into a respectable little article likely to be accepted for publication in a professional journal. For sociologists this means some little empirical study of a narrowly confined topic. In most instances such studies will require the application of statistical techniques. Since most professional journals in the field are suspicious of articles that do not contain at least some statistical material, this tendency is further strengthened. And so eager young sociologists stranded somewhere in hinterland institutions, yearning for the richer pastures of the better universities, supply us with a steady stream of little statistical studies of the dating habits of their students, the political opinions of the surrounding natives or the class system of some hamlet within commuting distance of their college. It might be added here that this system is not quite so terrible as it may seem to the newcomer to the field, since its ritual requirements are well known to all concerned. As a result, the sensible person reads the sociological journals mainly for the book reviews and the obituaries, and goes to sociological meetings only if he is looking for a job or has other intrigues to carry on.

The prominence of statistical techniques in American sociology today has, then, certain ritual functions that are readily understandable in view of the power system within which most sociologists have to make a career. In fact, most sociologists have little more than a cookbook knowledge of statistics, treating it with about the same mixture of awe,

21

ignorance and timid manipulation as a poor village priest would the mighty Latin cadences of Thomist theology. Once one has realized these things, however, it should be clear that sociology ought not to be judged by these aberrations. One then becomes, as it were, sociologically sophisticated about sociology, and enabled to look beyond the outward signs to whatever inward grace may be hidden behind them.

Statistical data by themselves do not make sociology. They become sociology only when they are sociologically interpreted, put within a theoretical frame of reference that is sociological. Simple counting, or even correlating different items that one counts, is not sociology. There is almost no sociology in the Kinsey reports. This does not mean that the data in these studies are not true or that they cannot be relevant to sociological understanding. They are, taken by themselves, raw materials that can be used in sociological interpretation. The interpretation, however, must be broader than the data themselves. So the sociologist cannot arrest himself at the frequency tables of premarital petting or extramarital pederasty. These enumerations are meaningful to him only in terms of their much broader implications for an understanding of institutions and values in our society. To arrive at such understanding the sociologist will often have to apply statistical techniques, especially when he is dealing with the mass phenomena of modern social life. But sociology consists of statistics as little as philology consists of conjugating irregular verbs or chemistry of making nasty smells in test tubes.

Another image of the sociologist current today and rather closely related to that of statistician is the one that sees him as a man mainly concerned in developing a scientific methodology that he can then impose on human phenomena. This image is frequently held by people in the humanities and presented as proof that sociology is a form of intellectual barbarism. One part of this criticism of sociology by the *littérateurs* is often a scathing commentary on the outlandish jargon in which much sociological writing is couched. By contrast, of course, the one who makes these criticisms offers himself as a guardian of the classical traditions of humane learning.

It would be quite possible to meet such criticism by an argument *ad hominem*. Intellectual barbarism seems to be fairly evenly distributed in the main scholarly disciplines dealing with the phenomenon 'man'. However, it is undignified to argue *ad hominem*, so we shall readily admit that, indeed, there is much that passes today under the heading of sociology that is justly called barbarian, if that word is intended to denote an ignorance of history and philosophy, narrow expertise without wider horizons, a preoccupation with technical skills, and total insensitivity to the uses of language. Once more, these elements can themselves be understood sociologically in terms of certain characteristics of contemporary academic life. The competition for prestige and jobs in fields rapidly becoming more and more complex forces specialization that all too frequently leads to a depressing parochialism of interests. But it would again be inaccurate to identify sociology with this much more pervasive intellectual trend.

Sociology has, from its beginnings, understood itself as a science. There has been much controversy about the precise meaning of this self-definition. For instance, German sociologists have emphasized the difference between the social and the natural sciences much more strongly than their French or American colleagues. But the allegiance of sociologists to the scientific ethos has meant everywhere a willingness to be bound by certain scientific canons of procedure. If the sociologist remains faithful to his calling, his statements must be arrived at through the observation of certain rules of evidence that allow others to check on or to repeat or to develop his findings further. It is this scientific discipline that often supplies the motive for reading a sociological work as against, say, a novel on the same topic that might describe matters in much more impressive and convincing language. As sociologists tried to develop their scientific rules of evidence, they were compelled to reflect upon methodological problems. This is why methodology is a necessary and valid part of the sociological enterprise.

At the same time it is quite true that some sociologists, especially in America, have become so preoccupied with methodological questions that they have ceased to be interested in society at all. As a result, they have found out nothing of significance about any aspect of social life, since in science as in love a concentration on technique is quite likely to lead to impotence. Much of this fixation on methodology can be explained in terms of the urge of a relatively new discipline to find acceptance on the academic scene. Since science is an almost sacred entity among Americans in general and American academicians in particular, the desire to emulate the procedures of the older natural sciences is very strong among the newcomers in the market-place of erudition. Giving in to this desire, the experimental psychologists, for instance, have succeeded to such an extent that their studies have commonly nothing more to do with anything that human beings are or do. The irony of this process lies in the fact that natural scientists themselves have been giving up the very positivistic dogmatism that their emulators are still straining to adopt. But this is not our concern here. Suffice it to say that sociologists have succeeded in avoiding some of the more grotesque exaggerations of this 'methodism', as compared with some fields close by. As they become more secure in their academic status, it may be expected that this methodological inferiority complex will diminish even further.

The charge that many sociologists write in a barbaric dialect must also be admitted with similar reservations. Any scientific discipline must develop a terminology. This is self-evident for a discipline such as, say, nuclear physics that deals with matters unknown to most people and for which no words exist in common speech. However, terminology is possibly even more important for the social sciences, just because their subject matter *is* familiar and just because words *do* exist to denote it. Because we are well acquainted with the social institutions that surround us, our perception of them is imprecise and often erroneous. In very much the same way most of us will have considerable difficulty giving an accurate description of our parents, husbands or wives, children or close friends. Also, our language is often (and perhaps blessedly) vague and confusing in its references to social reality. Take for an example the concept of *class*, a very important one in sociology. There must be dozens of meanings that this term may have in common speech – income brackets, races, ethnic groups, power cliques, intelligence ratings, and many others. It is obvious that the sociologist must have a precise, unambiguous definition of the concept if his work is to proceed with any degree of scientific rigour. In view of these facts, one can understand that some sociologists have been tempted to invent altogether new words to avoid the semantic traps of the vernacular usage. We would contend, then, that some of these neologisms have been necessary. We would also contend, however, that most sociology can be presented in intelligible English with but a little effort and that a good deal of contemporary 'sociologese' can be understood as a self-conscious mystification. Here again, however, we are confronted with an intellectual phenomenon that affects other fields as well. There may be a connexion with the strong influence of German academic life in a formative period in the development of American universities. Scientific profundity was gauged by the ponderousness of scientific language. If scientific prose was unintelligible to any but the narrow circle of initiates to the field in question, this was *ipso facto* proof of its intellectual respectability. Much American scholarly writing still reads like a translation from the German. This is certainly regrettable. It has little to do, however, with the legitimacy of the sociological enterprise as such.

Finally, we would look at an image of the sociologist not so much in his professional role as in his being, supposedly, a certain kind of person. This is the image of the sociologist as a detached, sardonic observer, and a cold manipulator of men. Where this image prevails, it may represent an ironic triumph of the sociologist's own efforts to be accepted as a genuine scientist. The sociologist here becomes the self-appointed superior man, standing off from the warm vitality of common existence, finding his satisfactions not in living but in coolly appraising the lives of others, filing them away in little categories, and thus presumably missing the real significance of what he is observing. Further, there is the notion that, when he involves himself in social processes at all, the sociologist does so as an uncommitted technician, putting his manipulative skills at the disposal of the powers that be.

This last image is probably not very widely held. It is mainly held by people concerned for political reasons with actual or possible misuses of sociology in modern societies. There is not very much to say about this image by way of refutation. As a general portrait of the contemporary sociologist it is certainly a gross distortion. It fits very few individuals that anyone is likely to meet in America today. The problem of the political role of the social scientist is, nevertheless, a very genuine one. For instance, the employment of sociologists by certain branches of industry and government raises moral questions that ought to be faced

more widely than they have been so far. These are, however, moral questions that concern all men in positions of
responsibility in modern society. The image of the sociologist as an observer without compassion and a manipulator
without conscience need not detain us further here. By and
large, history produces very few Talleyrands. As for contemporary sociologists, most of them would lack the
emotional equipment for such a role, even if they should
aspire to it in moments of feverish fantasy.

How then are we to conceive of the sociologist? In discussing the various images of him that abound in the popular
mind we have already brought out certain elements that
would have to go into our conception. We can now put
them together. In doing so, we shall construct what sociologists themselves call an 'ideal type'. This means that what
we delineate will not be found in reality in its pure form.
Instead, one will find approximations to it and deviations
from it, in varying degrees. Nor is it to be understood as an
empirical average. We would not even claim that all
individuals who now call themselves sociologists will recognize themselves without reservations in our conception, nor
would we dispute the right of those who do not so recognize
themselves to use the appellation. Our business is not
excommunication. We would, however, contend that our
'ideal type' corresponds to the self-conception of most
sociologists in the mainstream of the discipline, both
historically (at least in this century) and today.

The sociologist, then, is someone concerned with understanding society in a disciplined way. The nature of this
discipline is scientific. This means that what the sociologist
finds and says about the social phenomena he studies occurs
within a certain rather strictly defined frame of reference.
One of the main characteristics of this scientific frame of
reference is that operations are bound by certain rules of
evidence. As a scientist, the sociologist tries to be objective,

to control his personal preferences and prejudices, to perceive clearly rather than to judge normatively. This restraint, of course, does not embrace the totality of the sociologist's existence as a human being, but is limited to his operations *qua* sociologist. Nor does the sociologist claim that his frame of reference is the only one within which society can be looked at. For that matter, very few scientists in any field would claim today that one should look at the world only scientifically. The botanist looking at a daffodil has no reason to dispute the right of the poet to look at the same object in a very different manner. There are many ways of playing. The point is not that one denies other people's games but that one is clear about the rules of one's own. The game of the sociologist, then, uses scientific rules. As a result, the sociologist must be clear in his own mind as to the meaning of these rules. That is, he must concern himself with methodological questions. Methodology does not constitute his goal. The latter, let us recall once more, is the attempt to understand society. Methodology helps in reaching this goal. In order to understand society, or that segment of it that he is studying at the moment, the sociologist will use a variety of means. Among these are statistical techniques. Statistics can be very useful in answering certain sociological questions. But statistics does not constitute sociology. As a scientist, the sociologist will have to be concerned with the exact significance of the terms he is using. That is, he will have to be careful about terminology. This does not have to mean that he must invent a new language of his own, but it does mean that he cannot naïvely use the language of everyday discourse. Finally, the interest of the sociologist is primarily theoretical. That is, he is interested in understanding for its own sake. He may be aware of or even concerned with the practical applicability and consequences of his findings, but at that point he leaves the sociological frame of reference as such and moves into

28

realms of values, beliefs and ideas that he shares with other men who are not sociologists.

We daresay that this conception of the sociologist would meet with very wide consensus within the discipline today. But we would like to go a little bit farther here and ask a somewhat more personal (and therefore, no doubt, more controversial) question. We would like to ask not only what it is that the sociologist is doing but also what it is that drives him to it. Or, to use the phrase Max Weber used in a similar connexion, we want to inquire a little into the nature of the sociologist's demon. In doing so, we shall evoke an image that is not so much ideal-typical in the above sense but more confessional in the sense of personal commitment. Again, we are not interested in excommunicating anyone. The game of sociology goes on in a spacious playground. We are just describing a little more closely those we would like to tempt to join our game.

We would say then that the sociologist (that is, the one we would really like to invite to our game) is a person intensively, endlessly, shamelessly interested in the doings of men. His natural habitat is all the human gathering places of the world, wherever men come together. The sociologist may be interested in many other things. But his consuming interest remains in the world of men, their institutions, their history, their passions. And since he is interested in men, nothing that men do can be altogether tedious for him. He will naturally be interested in the events that engage men's ultimate beliefs, their moments of tragedy and grandeur and ecstasy. But he will also be fascinated by the commonplace, the everyday. He will know reverence, but this reverence will not prevent him from wanting to see and to understand. He may sometimes feel revulsion or contempt. But this also will not deter him from wanting to have his questions answered. The sociologist, in his quest for understanding, moves through the world of men without respect for the usual lines of demarcation. Nobility and degradation, power and obscurity, intelligence and folly – these are equally *interesting* to him, however unequal they may be in his personal values or tastes. Thus his questions may lead him to all possible levels of society, the best and the least known places, the most respected and the most despised. And, if he is a good sociologist, he will find himself in all these places because his own questions have so taken possession of him that he has little choice but to seek for answers.

It would be possible to say the same things in a lower key. We could say that the sociologist, but for the grace of his academic title, is the man who must listen to gossip despite himself, who is tempted to look through keyholes, to read other people's mail, to open closed cabinets. Before some otherwise unoccupied psychologist sets out now to construct an aptitude test for sociologists on the basis of sublimated voyeurism, let us quickly say that we are speaking merely by way of analogy. Perhaps some little boys consumed with curiosity to watch their maiden aunts in the bathroom later become inveterate sociologists. This is quite uninteresting. What interests us is the curiosity that grips any sociologist in front of a closed door behind which there are human voices. If he is a good sociologist, he will want to open that door, to understand these voices. Behind each closed door he will anticipate some new facet of human life not yet perceived and understood.

The sociologist will occupy himself with matters that others regard as too sacred or as too distasteful for dispassionate investigation. He will find rewarding the company of priests or of prostitutes, depending not on his personal preferences but on the questions he happens to be asking at the moment. He will also concern himself with matters that others may find much too boring. He will be interested in the human interaction that goes with warfare or with great intellectual discoveries, but also in the relations between people employed in a restaurant or between a group of little girls playing with their dolls. His main focus of attention is not the ultimate significance of what men do, but the action in itself, as another example of the infinite richness of human conduct. So much for the image of our playmate.

In these journeys through the world of men the sociologist will inevitably encounter other professional Peeping Toms. Sometimes these will resent his presence, feeling that he is poaching on their preserves. In some places the sociologist will meet up with the economist, in others with the political scientist, in yet others with the psychologist or the ethnologist. Yet chances are that the questions that have brought him to these same places are different from the ones that propelled his fellow-trespassers. The sociologist's questions always remain essentially the same: 'What are people doing with each other here?' 'What are their relationships to each other?' 'How are those relationships organized in institutions?' 'What are the collective ideas that move men and institutions?' In trying to answer these questions in specific instances, the sociologist will, of course, have to deal with economic or political matters, but he will do so in a way rather different from that of the economist or the political scientist. The scene that he contemplates is the same human scene that these other scientists concern themselves with. But the sociologist's angle of vision is different. When this is understood, it becomes clear that it makes little sense to try to stake out a special enclave within which the sociologist will carry on business in his own right. Like Wesley, the sociologist will have to confess that his parish is the world. But unlike some latter-day Wesleyans he will gladly share this parish with others. There is, however, one traveller whose path the sociologist will cross more often than anyone else's on his journeys. This is the historian. Indeed, as soon as the sociologist turns from the present to the past, his preoccupations are very hard indeed to distinguish from those of the historian. However, we shall leave this relationship to a later part of our considerations. Suffice it to say here that the sociological journey will be much impoverished unless it is punctuated frequently by conversation with that other particular traveller.

Any intellectual activity derives excitement from the moment it becomes a trail of discovery. In some fields of learning this is the discovery of worlds previously unthought and unthinkable. This is the excitement of the astronomer or of the nuclear physicist on the antipodal boundaries of the realities that man is capable of conceiving. But it can also be the excitement of bacteriology or geology. In a different way it can be the excitement of the linguist discovering new realms of human expression or of the anthropologist exploring human customs in faraway countries. In such discovery, when undertaken with passion, a widening of awareness, sometimes a veritable transformation of consciousness, occurs. The universe turns out to be much more wonder-full than one had ever dreamed. The excitement of sociology is usually of a different sort. Sometimes, it is true, the sociologist penetrates into worlds that had previously been quite unknown to him – for instance, the world of crime, or the world of some bizarre religious sect, or the world fashioned by the exclusive concerns of some group such as medical specialists or military leaders or advertising executives. However, much of the time the sociologist moves in sectors of experience that are familiar to him and to most people in his society. He investigates communities, institutions and activities that one can read about every day in the newspapers. Yet there is another excitement of discovery beckoning in his investigations. It is not the excitement of coming upon the totally unfamiliar, but rather the excitement of finding the familiar becoming transformed in its meaning. The fascination of sociology lies in the fact that its perspective makes us see in a new light the very world in which we have lived all our lives. This also constitutes a transformation of consciousness. Moreover, this transformation is more relevant existentially than that of many other intellectual disciplines, because it is more difficult to segregate in some special compartment of the mind. The astronomer does not live in the remote galaxies, and the nuclear physicist can, outside his laboratory, eat and laugh and marry and vote without thinking about the insides of the atom. The geologist looks at rocks only at appropriate times, and the linguist speaks English with his wife. The sociologist lives in society, on the job and off it. His own life, inevitably, is part of his subject matter. Men being what they are, sociologists too manage to segregate their professional insights from their everyday affairs. But it is a rather difficult feat to perform in good faith.

The sociologist moves in the common world of men, close to what most of them would call real. The categories he employs in his analyses are only refinements of the categories by which other men live – power, class, status, race, ethnicity. As a result, there is a deceptive simplicity and obviousness about some sociological investigations. One reads them, nods at the familiar scene, remarks that one has heard all this before and don't people have better things to do than to waste their time on truisms – until one is suddenly brought up against an insight that radically questions everything one had previously assumed about this familiar scene. This is the point at which one begins to sense the excitement of sociology.

Let us take a specific example. Imagine a sociology class in a Southern college where almost all the students are white Southerners. Imagine a lecture on the subject of the racial system of the South. The lecturer is talking here of matters that have been familiar to his students from the time of their infancy. Indeed, it may be that they are much more familiar with the minutiae of this system than he is. They are quite bored as a result. It seems to them that he is only using more pretentious words to describe what they already know. Thus he may use the term 'caste', one commonly used now by American sociologists to describe the Southern racial system. But in explaining the term he shifts to traditional Hindu society, to make it clearer. He then goes on to analyse the magical beliefs inherent in caste taboos, the social dynamics of commensalism and connubium, the economic interests concealed within the system, the way in which religious beliefs relate to the taboos, the effects of the caste system upon the industrial development of the society and vice versa – all in India. But suddenly India is not very far away at all. The lecture then goes back to its Southern theme. The familiar now seems not quite so familiar any more. Questions are raised that are new, perhaps raised angrily, but raised all the same. And at least some of the students have begun to understand that there are functions involved in this business of race that they have not read about in the newspapers (at least not those in their home-towns) and that their parents have not told them – partly, at least, because neither the newspapers nor the parents knew about them.

It can be said that the first wisdom of sociology is this – things are not what they seem. This too is a deceptively simple statement. It ceases to be simple after a while. Social reality turns out to have many layers of meaning. The discovery of each new layer changes the perception of the whole.

Anthropologists use the term 'culture shock' to describe the impact of a totally new culture upon a newcomer. In an extreme instance such shock will be experienced by the Western explorer who is told, halfway through dinner, that he is eating the nice old lady he had been chatting with the previous day – a shock with predictable physiological if not moral consequences. Most explorers no longer encounter cannibalism in their travels today. However, the first encounters with polygamy or with puberty rites or even with the way some nations drive their automobiles can be quite a shock to an American visitor. With the shock may go not only disapproval or disgust but a sense of excitement that things can *really* be that different from what they are at home. To some extent, at least, this is the excitement of any first travel abroad. The experience of sociological discovery could be described as 'culture shock' minus geographical displacement. In other words, the sociologist travels at home – with shocking results. He is unlikely to find that he is eating a nice old lady for dinner. But the discovery, for instance, that his own church has considerable money invested in the missile industry or that a few blocks from his home there are people who engage in cultic orgies may not be drastically different in emotional impact. Yet we would not want to imply that sociological discoveries are always or even usually outrageous to moral sentiment. Not at all. What they have in common with exploration in distant lands, however, is the sudden illumination of new and unsuspected facets of human existence in society. This is the excitement and, as we shall try to show later, the humanistic justification of sociology.

People who like to avoid shocking discoveries, who prefer to believe that society is just what they were taught in Sunday School, who like the safety of the rules and the maxims of what Alfred Schuetz has called the 'world-taken-for-granted', should stay away from sociology. People who feel no temptation before closed doors, who have no curiosity about human beings, who are content to admire scenery without wondering about the people who live in those houses on the other side of that river, should probably also stay away from sociology. They will find it unpleasant or, at any rate, unrewarding. People who are interested in human beings only if they can change, convert or reform them should also be warned, for they will find sociology much less useful than they hoped. And people whose interest is mainly in their own conceptual constructions will do just as well to turn to the study of little white mice. Sociology will be satisfying, in the long run, only to those who can think of nothing more entrancing than to watch men and to understand things human.

It may now be clear that we have, albeit deliberately, understated the case in the title of this chapter. To be sure, sociology is an individual pastime in the sense that it interests some men and bores others. Some like to observe human beings, others to experiment with mice. The world is big enough to hold all kinds and there is no logical priority for one interest as against another. But the word 'pastime' is weak in describing what we mean. Sociology is more like a passion. The sociological perspective is more like a demon that possesses one, that drives one compellingly, again and again, to the questions that are its own. An introduction to sociology is, therefore, an invitation to a very special kind of passion. No passion is without its dangers. The sociologist who sells his wares should make sure that he clearly pronounces a *caveat emptor* quite early in the transaction.

2

SOCIOLOGY AS A FORM OF CONSCIOUSNESS

IF the previous chapter has been successful in its presentation, it will be possible to accept sociology as an intellectual preoccupation of interest to certain individuals. To stop at this point, however, would in itself be very unsociological indeed. The very fact that sociology appeared as a discipline at a certain stage of Western history should compel us to ask further how it is possible for certain individuals to occupy themselves with it and what the preconditions are for this occupation. In other words, sociology is neither a timeless nor a necessary undertaking of the human mind. If this is conceded, the question logically arises as to the timely factors that made it a necessity to specific men. Perhaps, indeed, no intellectual enterprise is timeless or necessary. But religion, for instance, has been well-nigh universal in provoking intensive mental preoccupation throughout human history, while thoughts designed to solve the economic problems of existence have been a necessity in most human cultures. Certainly this does not mean that theology or economics, in our contemporary sense, are universally present phenomena of the mind, but we are at least on safe ground if we say that there always seems to have been human thought directed towards the problems that now constitute the subject matter of these disciplines. Not even this much, however, can be said of sociology. It presents itself rather as a peculiarly modern and Western cogitation. And, as we shall try to argue in this chapter, it is constituted by a peculiarly modern form of consciousness.

The peculiarity of sociological perspective becomes clear with some reflection concerning the meaning of the term 'society', a term that refers to the object *par excellence* of the discipline. Like most terms used by sociologists, this one is derived from common usage, where its meaning is imprecise. Sometimes it means a particular band of people (as in 'Society for the Prevention of Cruelty to Animals'), sometimes only those people endowed with great prestige or privilege (as in 'Boston society ladies'), and on other occasions it is simply used to denote company of any sort (for example, 'he greatly suffered in those years for lack of society'). There are other, less frequent meanings as well. The sociologist uses the term in a more precise sense, though, of course, there are differences in usage within the discipline itself. The sociologist thinks of 'society' as denoting a large complex of human relationships, or to put it in more technical language, as referring to a system of interaction. The word 'large' is difficult to specify quantitatively in this context. The sociologist may speak of a 'society' including millions of human beings (say, 'American society'), but he may also use the term to refer to a numerically much smaller collectivity (say, 'the society of second-year students here'). Two people chatting on a street corner will hardly constitute a 'society', but three people stranded on an island certainly will. The applicability of the concept, then, cannot be decided on quantitative grounds alone. It rather applies when a complex of relationships is sufficiently succinct to be analysed by itself, understood as an autonomous entity, set against others of the same kind.

The adjective 'social' must be similarly sharpened for sociological use. In common speech it may denote, once more, a number of different things – the informal quality of a certain gathering ('this is a social meeting – let's not discuss business'), an altruistic attitude on somebody's part ('he had a strong social concern in his job'), or, more generally, anything derived from contact with other people ('a social disease'). The sociologist will use the term more

narrowly and more precisely to refer to the quality of interaction, inter-relationship, mutuality. Thus two men chatting
on a street corner do not constitute a 'society', but what
transpires between them is certainly 'social'. 'Society' consists of a complex of such 'social' events. As to the exact
definition of the 'social', it is difficult to improve on Max
Weber's definition of a 'social' situation as one in which
people orient their actions towards one another. The web
of meanings, expectations and conduct resulting from such
mutual orientation is the stuff of sociological analysis.

Yet this refinement of terminology is not enough to show
up the distinctiveness of the sociological angle of vision.
We may get closer by comparing the latter with the perspective of other disciplines concerned with human actions.
The economist, for example, is concerned with the analyses
of processes that occur in society and that can be described
as social. These processes have to do with the basic problem of economic activity – the allocation of scarce goods
and services within a society. The economist will be concerned with these processes in terms of the way in which
they carry out, or fail to carry out, this function. The
sociologist, in looking at the same processes, will naturally
have to take into consideration their economic purpose. But
his distinctive interest is not necessarily related to this purpose as such. He will be interested in a variety of human
relationships and interactions that may occur here and that
may be quite irrelevant to the economic goals in question.
Thus economic activity involves relationships of power,
prestige, prejudice or even play that can be analysed with
only marginal reference to the properly economic function
of the activity.

The sociologist finds his subject matter present in all
human activities, but not all aspects of these activities constitute this subject matter. Social interaction is not some
specialized sector of what men do with each other. It is

rather a certain aspect of all these doings. Another way of putting this is by saying that the sociologist carries on a special sort of abstraction. The social, as an object of inquiry, is not a segregated field of human activity. Rather (to borrow a phrase from Lutheran sacramental theology) it is present 'in, with and under' many different fields of such activity. The sociologist does not look at phenomena that nobody else is aware of. But he looks at the same phenomena in a different way.

As a further example we could take the perspective of the lawyer. Here we actually find a point of view much broader in scope than that of the economist. Almost any human activity can, at one time or another, fall within the province of the lawyer. This, indeed, is the fascination of the law. Again, we find here a very special procedure of abstraction. From the immense wealth and variety of human deportment the lawyer selects those aspects that are pertinent (or, as he would say, 'material') to his very particular frame of reference. As anyone who has ever been involved in a lawsuit well knows, the criteria of what is relevant or irrelevant legally will often greatly surprise the principals in the case in question. This need not concern us here. We would rather observe that the legal frame of reference consists of a number of carefully defined models of human activity. Thus we have clear models of obligation, responsibility or wrong-doing. Definite conditions have to prevail before any empirical act can be subsumed under one of these headings, and these conditions are laid down by statutes or precedent. When these conditions are not met, the act in question is legally irrelevant. The expertise of the lawyer consists of knowing the rules by which these models are constructed. He knows, within his frame of reference, when a business contract is binding, when the driver of an automobile may be held to be negligent, or when rape has taken place.

The sociologist may look at these same phenomena, but his frame of reference will be quite different. Most importantly, his perspective on these phenomena cannot be derived from statutes or precedent. His interest in the human relationships occurring in a business transaction has no bearing on the legal validity of contracts signed, just as sociologically interesting deviance in sexual behaviour may not be capable of being subsumed under some particular legal heading. From the lawyer's point of view, the sociologist's inquiry is extraneous to the legal frame of reference. One might say that, with reference to the conceptual edifice of the law, the sociologist's activity is subterranean in character. The lawyer is concerned with what may be called the official conception of the situation. The sociologist often deals with very unofficial conceptions indeed. For the lawyer the essential thing to understand is how the law looks upon a certain type of criminal. For the sociologist it is equally important to see how the criminal looks at the law.

To ask sociological questions, then, presupposes that one is interested in looking some distance beyond the commonly accepted or officially defined goals of human actions. It presupposes a certain awareness that human events have different levels of meaning, some of which are hidden from the consciousness of everyday life. It may even presuppose a measure of suspicion about the way in which human events are officially interpreted by the authorities, be they political, juridical or religious in character. If one is willing to go as far as that, it would seem evident that not all historical circumstances are equally favourable for the development of sociological perspective.

It would appear plausible, in consequence, that sociological thought would have the best chance to develop in historical circumstances marked by severe jolts to the self-conception, especially the official and authoritative and generally accepted self-conception, of a culture. It is only in such circumstances that perceptive men are likely to be motivated to think beyond the assertions of this self-conception and, as a result, question the authorities. Albert Salomon has argued cogently that the concept of 'society', in its modern sociological sense, could emerge only as the normative structures of Christendom and later of the *ancien régime* were collapsing. We can, then, again conceive of 'society' as the hidden fabric of an edifice, the outside façade of which hides that fabric from the common view. In medieval Christendom, 'society' was rendered invisible by the imposing religio-political façade that constituted the common world of European man. As Salomon pointed out, the more secular political façade of the absolute state performed the same function after the Reformation had broken up the unity of Christendom. It was with the disintegration of the absolute state that the underlying frame of 'society' came into view – that is, a world of motives and forces that could not be understood in terms of the official interpretations of social reality. Sociological perspective can then be understood in terms of such phrases as 'seeing through', 'looking behind', very much as such phrases would be employed in common speech – 'seeing through his game', 'looking behind the scenes' – in other words, 'being up on all the tricks'.

We will not be far off if we see sociological thought as part of what Nietzsche called 'the art of mistrust'. Now, it would be a gross oversimplification to think that this art has existed only in modern times. 'Seeing through' things is probably a pretty general function of intelligence, even in very primitive societies. The American anthropologist Paul Radin has provided us with a vivid description of the sceptic as a human type in primitive culture. We also have evidence from civilizations other than that of the modern West, bearing witness to forms of consciousness that could well be called proto-sociological. We could point, for instance, to Herodotus or to Ibn-Khaldun. There are even texts from ancient Egypt evincing a profound disenchantment with a political and social order that has acquired the reputation of having been one of the most cohesive in human history. However, with the beginning of the modern era in the West this form of consciousness intensifies, becomes concentrated and systematized, marks the thought of an increasing number of perceptive men. This is not the place to discuss in detail the prehistory of sociological thought, a discussion in which we owe very much to Salomon. Nor would we even give here an intellectual table of ancestors for sociology, showing its connexions with Machiavelli, Erasmus, Bacon, seventeenth century philosophy and eighteenth century *belles-lettres* – this has been done elsewhere and by others much more qualified than this writer. Suffice it to stress once more that sociological thought marks the fruition of a number of intellectual developments that have a very specific location in modern Western history.

Let us return instead to the proposition that sociological perspective involves a process of 'seeing through' the façades of social structures. We could think of this in terms of a common experience of people living in large cities. One of the fascinations of a large city is the immense variety of human activities taking place behind the seemingly anonymous and endlessly undifferentiated rows of houses. A person who lives in such a city will time and again experience surprise or even shock as he discovers the strange pursuits that some men engage in quite unobtrusively in houses that, from the outside, look like all the others on a certain street. Having had this experience once or twice, one will repeatedly find oneself walking down a street, perhaps late in the evening, and wondering what may be going on under the bright lights showing through a line of drawn curtains. An ordinary family engaged in pleasant talk with guests? A scene of desperation amid illness or death? Or a scene of debauched pleasures? Perhaps a strange cult or a dangerous conspiracy? The façades of the houses cannot tell us, proclaiming nothing but an architectural conformity to the tastes of some group or class that may not even inhabit the street any longer. The social mysteries lie behind the façades. The wish to penetrate to these mysteries is an analogon to sociological curiosity. In some cities that are suddenly struck by calamity this wish may be abruptly realized. Those who have experienced wartime bombings know of the sudden encounters with unsuspected (and sometimes unimaginable) fellow tenants in the air-raid shelter of one's apartment building. Or they can recollect the startling morning sight of a house hit by a bomb during the night, neatly sliced in half, the façade torn away and the previously hidden interior mercilessly revealed in the daylight. But in most cities that one may normally live in, the façades must be penetrated by one's own inquisitive intrusions. Similarly, there are historical situations in which the façades of society are violently torn apart and all but the most incurious are forced to see that there was a reality behind the façades all along. Usually this does not happen and the façades continue to confront us with seemingly rock-like permanence. The perception of the reality behind the façades then demands a considerable intellectual effort.

A few examples of the way in which sociology 'looks behind' the façades of social structures might serve to make our argument clearer. Take, for instance, the political organization of a community. If one wants to find out how a modern American city is governed, it is very easy to get the official information about this subject. The city will have a charter, operating under the laws of the state. With some advice from informed individuals, one may look up various statutes that define the constitution of the city. Thus one may find out that this particular community has a city-

manager form of administration, or that party affiliations
do not appear on the ballot in municipal elections, or that
the city government participates in a regional water district.
In similar fashion, with the help of some newspaper reading, one may find out the officially recognized political
problems of the community. One may read that the city
plans to annex a certain suburban area, or that there has
been a change in the zoning ordinances to facilitate industrial development in another area, or even that one of the
members of the city council has been accused of using his
office for personal gain. All such matters still occur on the,
as is were, visible, official or public level of political life.
However, it would be an exceedingly naïve person who
would believe that this kind of information gives him a
rounded picture of the political reality of that community.
The sociologist will want to know above all the constituency
of the 'informal power structure' (as it has been called by
Floyd Hunter, an American sociologist interested in such
studies), which is a configuration of men and their power
that cannot be found in any statutes, and probably cannot
be read about in the newspapers. The political scientist or
the legal expert might find it very interesting to compare the
city charter with the constitutions of other similar communities. The sociologist will be far more concerned with
discovering the way in which powerful vested interests
influence or even control the actions of officials elected
under the charter. These vested interests will not be found
in city hall, but rather in the executive suites of corporations
that may not even be located in that community, in the
private mansions of a handful of powerful men, perhaps
in the offices of certain labour unions or even, in some
instances, in the headquarters of criminal organizations.
When the sociologist concerns himself with power, he will
'look behind' the official mechanisms that are supposed to
regulate power in the community. This does not necessarily

mean that he will regard the official mechanisms as totally ineffective or their legal definition as totally illusionary. But at the very least he will insist that there is another level of reality to be investigated in the particular system of power. In some cases he might conclude that to look for real power in the publicly recognized places is quite delusional.

Take another example. Protestant denominations in America differ widely in their so-called 'polity', that is, the officially defined way in which the denomination is run. One may speak of an episcopal, a presbyterian or a congregational 'polity' (meaning by this not the denominations called by these names, but the forms of ecclesiastical government that various denominations share – for instance, the episcopal form shared by Episcopalians and Methodists, the congregational by Congregationalists and Baptists). In nearly all cases, the 'polity' of a denomination is the result of a long historical development and is based on a theological rationale over which the doctrinal experts continue to quarrel. Yet a sociologist interested in studying the government of American denominations would do well not to arrest himself too long at these official definitions. He will soon find that the real questions of power and organization have little to do with 'polity' in the theological sense. He will discover that the basic form of organization in all denominations of any size is bureaucratic. The logic of administrative behaviour is determined by bureaucratic processes, only very rarely by the workings of an episcopal or a congregational point of view. The sociological investigator will then quickly 'see through' the mass of confusing terminology denoting office-holders in the ecclesiastical bureaucracy and correctly identify those who hold executive power, no matter whether they be called 'bishops', or 'stated clerks' or 'synod presidents'. Understanding denominational organization as belonging to the much larger species of bureaucracy, the sociologist will then be able to grasp the processes that occur in the organization, to observe the internal and external pressures brought to bear on those who are theoretically in charge. In other words, behind the façade of an 'episcopal polity' the sociologist will perceive the workings of a bureaucratic apparatus that is not terribly different in the Methodist Church, an agency of the Federal government, General Motors or the United Automobile Workers.

Or take an example from economic life. The personnel manager of an industrial plant will take delight in preparing brightly coloured charts that show the table of organization that is supposed to administer the production process. Every man has his place, every person in the organization knows from whom he receives his orders and to whom he must transmit them, every work team has its assigned role in the great drama of production. In reality things rarely work this way – and every good personnel manager knows this. Superimposed on the official blueprint of the organization is a much subtler, much less visible network of human groups, with their loyalties, prejudices, antipathies and (most important) codes of behaviour. Industrial sociology is full of data on the operations of this informal network, which always exists in varying degrees of accommodation and conflict with the official system. Very much the same coexistence of formal and informal organization are to be found wherever large numbers of men work together or live together under a system of discipline – military organizations, prisons, hospitals, schools, going back to the mysterious leagues that children form among themselves and that their parents only rarely discern. Once more, the sociologist will seek to penetrate the smoke screen of the official versions of reality (those of the foreman, the officer, the teacher) and try to grasp the signals that come from the 'underworld' (those of the worker, the enlisted man, the schoolboy).

Let us take one further example. In Western countries, and especially in America, it is assumed that men and women marry because they are in love. There is a broadly based popular mythology about the character of love as a violent, irresistible emotion that strikes where it will, a mystery that is the goal of most young people and often of the not-so-young as well. As soon as one investigates, however, which people actually marry each other, one finds that the lightning-shaft of Cupid seems to be guided rather strongly within very definite channels of class, income, education, racial and religious background. If one then investigates a little further into the behaviour that is engaged in prior to marriage under the rather misleading euphemism of 'courtship', one finds channels of interaction that are often rigid to the point of ritual. The suspicion begins to dawn on one that, most of the time, it is not so much the emotion of love that creates a certain kind of relationship, but that carefully predefined and often planned relationships eventually generate the desired emotion. In other words, when certain conditions are met or have been constructed, one allows oneself 'to fall in love'. The sociologist investigating our patterns of 'courtship' and marriage soon discovers a complex web of motives related in many ways to the entire institutional structure within which an individual lives his life – class, career, economic ambition, aspirations of power and prestige. The miracle of love now begins to look somewhat synthetic. Again, this need not mean in any given instance that the sociologist will declare the romantic interpretation to be an illusion. But, once more, he will look beyond the immediately given and publicly approved interpretations. Contemplating a couple who in their turn are contemplating the moon, the sociologist need not feel constrained to deny the emotional impact of the scene thus illuminated. But he will observe the machinery that went into the construction of the scene in its non-lunar aspects – the status index of the automobile from which the contemplation occurs, the canons of taste and tactics that determine the costume of the contemplators, the many ways in which language and demeanour place them socially, thus the social location and intentionality of the entire enterprise.

It may have become clear at this point that the problems that will interest the sociologist are not necessarily what other people may call 'problems'. The way in which public officials and newspapers (and, alas, some college textbooks in sociology) speak about 'social problems' serves to obscure this fact. People commonly speak of a 'social problem' when something in society does not work the way it is supposed to according to the official interpretations. They then expect the sociologist to study the 'problem' as they have defined it and perhaps even to come up with a 'solution' that will take care of the matter to their own satisfaction. It is important, against this sort of expectation, to understand that a sociological problem is something quite different from a 'social problem' in this sense. For example, it is naïve to concentrate on crime as a 'problem' because law-enforcement agencies so define it, or on divorce because that is a 'problem' to the moralists of marriage. Even more clearly, the 'problem' of the foreman to get his men to work more efficiently or of the line officer to get his troops to charge the enemy more enthusiastically need not be problematic at all to the sociologist (leaving out of consideration for the moment the probable fact that the sociologist asked to study such 'problems' is employed by the corporation or the army). The sociological problem is always the understanding of what goes on here in terms of social interaction. Thus the sociological problem is not so much why some things 'go wrong' from the viewpoint of the authorities and the management of the social scene, but how the whole system works in the first place, what are its presuppositions and by what means it is held together. The fundamental sociological problem is not crime but the law, not divorce but marriage, not racial discrimination but racially defined stratification, not revolution but government.

This point can be explicated further by an example. Take a settlement house in a lower-class slum district trying to wean away teenagers from the publicly disapproved activities of a juvenile gang. The frame of reference within which social workers and police officers define the 'problems' of this situation is constituted by the world of middle-class, respectable, publicly approved values. It is a 'problem' if teenagers drive around in stolen automobiles, and it is a 'solution' if instead they will play group games in the settlement house. But if one changes the frame of reference and looks at the situation from the viewpoint of the leaders of the juvenile gang, the 'problems' are defined in reverse order. It is a 'problem' for the solidarity of the gang if its members are seduced away from those activities that lend prestige to the gang within its own social world, and it would be a 'solution' if the social workers went way the hell back uptown where they came from. What is a 'problem' to one social system is the normal routine of things to the other system, and vice versa. Loyalty and disloyalty, solidarity and deviance, are defined in contradictory terms by the representatives of the two systems. Now, the sociologist may, in terms of his own values, regard the world of middle-class respectability as more desirable and therefore want to come to the assistance of the settlement house, which is its missionary outpost *in partibus infidelium*. This, however, does not justify the identification of the director's headaches with what are 'problems' sociologically. The 'problems' that the sociologist will want to solve concern an understanding of the entire social situation, the values and modes of action in *both* systems, and the way in which the two systems coexist in space and time. Indeed, this very ability to look at a situation from the vantage points of competing systems of interpretation is, as we shall see more clearly later on, one of the hallmarks of sociological consciousness.

We would contend, then, that there is a debunking motif inherent in sociological consciousness. The sociologist will be driven time and again, by the very logic of his discipline, to debunk the social systems he is studying. This unmasking tendency need not necessarily be due to the sociologist's temperament or inclinations. Indeed, it may happen that the sociologist, who as an individual may be of a conciliatory disposition and quite disinclined to disturb the comfortable assumptions on which he rests his own social existence, is nevertheless compelled by what he is doing to fly in the face of what those around him take for granted. In other words, we would contend that the roots of the debunking motif in sociology are not psychological but methodological. The sociological frame of reference, with its built-in procedure of looking for levels of reality other than those given in the official interpretations of society, carries with it a logical imperative to unmask the pretensions and the propaganda by which men cloak their actions with each other. This unmasking imperative is one of the characteristics of sociology particularly at home in the temper of the modern era.

The debunking tendency in sociological thought can be illustrated by a variety of developments within the field. For example, one of the major themes in Weber's sociology is that of the unintended, unforeseen consequences of human actions in society. Weber's most famous work, *The Protestant Ethic and the Spirit of Capitalism*, in which he demonstrated the relationship between certain consequences of Protestant values and the development of the capitalist ethos, has often been misunderstood by critics precisely because they missed this theme. Such critics have pointed out that the Protestant thinkers quoted by Weber never intended their teachings to be applied so as to produce the specific economic results in question. Specifically, Weber argued that the Calvinist doctrine of predestination led people to behave in what he called an 'inner-worldly ascetic' way, that is, in a manner that concerns itself intensively, systematically and selflessly with the affairs of this world, especially with economic affairs. Weber's critics have then pointed out that nothing was farther from the mind of Calvin and the other leaders of the Calvinist Reformation. But Weber never maintained that Calvinist thought *intended* to produce these economic action patterns. On the contrary, he knew very well that the intentions were drastically different. The consequences took place regardless of intentions. In other words, Weber's work (and not only the famous part of it just mentioned) gives us a vivid picture of the *irony* of human actions. Weber's sociology thus provides us with a radical antithesis to any views that understand history as the realization of ideas or as the fruit of the deliberate efforts of individuals or collectivities. This does not mean at all that ideas are not important. It does mean that the outcome of ideas is commonly very different from what those who had the ideas in the first place planned or hoped. Such a consciousness of the ironic aspect of history is sobering, a strong antidote to all kinds of revolutionary utopianism.

The debunking tendency of sociology is implicit in all sociological theories that emphasize the autonomous character of social processes. For instance, Émile Durkheim, the founder of the most important school in French sociology, emphasized that society was a reality *sui generis*, that is, a reality that could not be reduced to psychological or other factors on different levels of analysis. The effect of this insistence has been a sovereign disregard for individually intended motives and meanings in Durkheim's study of various phenomena. This is perhaps most sharply revealed in his well-known study of suicide, in the work of that title, where individual intentions of those who commit or try to commit suicide are completely left out of the analysis in favour of statistics concerning various social characteristics of these individuals. In the Durkheimian perspective, to live in society means to exist under the domination of society's logic. Very often men act by this logic without knowing it. To discover this inner dynamic of society, therefore, the sociologist must frequently disregard the answers that the social actors themselves would give to his questions and look for explanations that are hidden from their own awareness. This essentially Durkheimian approach has been carried over into the theoretical approach now called functionalism. In functional analysis society is analyzed in terms of its own workings as a system, workings that are often obscure or opaque to those acting within the system. The contemporary American sociologist Robert Merton has expressed this approach well in his concepts of 'manifest' and 'latent' functions. The former are the conscious and deliberate functions of social processes, the latter the unconscious and unintended ones. Thus the 'manifest' function of anti-gambling legislation may be to suppress gambling, its 'latent' function to create an illegal empire for the gambling syndicates. Or Christian missions in parts of Africa 'manifestly' tried to convert Africans to Christianity, 'latently' helped to destroy the indigenous tribal cultures and thus provided an important impetus towards rapid social transformation. Or the control of the Communist Party over all sectors of social life in Russia 'manifestly' was to assure the continued dominance of the revolutionary ethos, 'latently' created a new class of comfortable bureaucrats uncannily bourgeois in its aspirations and increasingly disinclined toward the self-denial of Bolshevik dedication. Or the 'manifest' function of many voluntary associations in America is sociability and public service, the 'latent' function to attach status indices to those permitted to belong to such associations.

The concept of 'ideology', a central one in some sociological theories, could serve as another illustration of the debunking tendency discussed. Sociologists speak of 'ideology' in discussing views that serve to rationalize the vested interests of some group. Very frequently such views systematically distort social reality in much the same way that an individual may neurotically deny, deform or reinterpret aspects of his life that are inconvenient to him. The important approach of the Italian sociologist Vilfredo Pareto has a central place for this perspective and, as we shall see in a later chapter, the concept of 'ideology' is essential for the approach called the 'sociology of knowledge'. In such analyses the ideas by which men explain their actions are unmasked as self-deception, sales talk, the kind of 'sincerity' that David Riesman has aptly described as the state of mind of a man who habitually believes his own propaganda. In this way, we can speak of 'ideology' when we analyze the belief of many American physicians that standards of health will decline if the fee-for-service method of payment is abolished, or the conviction of many undertakers that inexpensive funerals show lack of affection for the departed, or the definition of their activity by quizmasters on television as 'education'. The self-image of the insurance salesman as a fatherly adviser to young families, of the burlesque stripper as an artist, of the propagandist as a communications expert, of the hangman as a public servant – all these notions are not only individual assuagements of guilt or status anxiety, but constitute the official self-interpretations of entire social groups, obligatory for their members on pain of excommunication. In uncovering the social functionality of ideological pretensions the sociologist will try not to resemble those historians of whom Marx said that every corner grocer is superior to them in knowing the difference between what a man is and what he claims to be. The debunking motif of sociology lies in this penetration of verbal smoke screens to the unadmitted and often unpleasant mainsprings of action.

It has been suggested above that sociological consciousness is likely to arise when the commonly accepted or authoritatively stated interpretations of society become shaky. As we have already said, there is a good case for thinking of the origins of sociology in France (the mother country of the discipline) in terms of an effort to cope intellectually with the consequences of the French Revolution, not only of the one great cataclysm of 1789 but of what De Tocqueville called the continuing Revolution of the nineteenth century. In the French case it is not difficult to perceive sociology against the background of the rapid transformations of modern society, the collapse of façades, the deflation of old creeds and the upsurge of frightening new forces on the social scene. In Germany, the other European country in which an important sociological movement arose in the nineteenth century, the matter has a rather different appearance. If one may quote Marx once more, the Germans had a tendency to carry on in professors' studies the revolutions that the French performed on the barricades. At least one of these academic roots of revolution, perhaps the most important one, may be sought in the broadly based movement of thought that came to be called 'historicism'. This is not the place to go into the full story of this movement. Suffice it to say that it represents an attempt to deal philosophically with the overwhelming sense of the relativity of all values in history. This awareness of relativity was an almost necessary outcome of the immense accumulation of German historical scholarship in every conceivable field. Sociological thought was at least partly grounded in the need to bring order and intelligibility to the impression of chaos that this array of historical knowledge made on some observers. Needless to stress, however, the society of the German sociologist was changing all around him just as was that of his French colleague, as Germany rushed towards industrial power and nationhood in the second half of the nineteenth century. We shall not pursue these questions, though. If we turn to America, the country in which sociology came to receive its most widespread acceptance, we find once more a different set of circumstances, though again against a background of rapid and profound social change. In looking at this American development we can detect another motif of sociology, closely related to that of debunking but not identical with it – its fascination with the unrespectable view of society.

In at least every Western society it is possible to distinguish between respectable and unrespectable sectors. In that respect American society is not in a unique position. But American respectability has a particularly pervasive quality about it. This may be ascribed in part, perhaps, to the lingering after-effects of the Puritan way of life. More probably it has to do with the predominant role played by the bourgeoisie in shaping American culture. Be this as it may in terms of historical causation, it is not difficult to look at social phenomena in America and place them readily in one of these two sectors. We can perceive the official, respectable America represented symbolically by the Chamber of Commerce, the churches, the schools and other centres of civic ritual. But facing this world of respectability is an 'other America', present in every town of any size, an America that has other symbols and that speaks another language. This language is probably its safest identification tag. It is the language of the poolroom and the poker game, of bars, brothels and army barracks. But it is also the language that breaks out with a sigh of relief between two salesmen having a drink in the parlour car as their train races past clean little Midwestern villages on a Sunday morning, with clean little villagers trooping into the whitewashed sanctuaries. It is the language that is suppressed in the company of ladies and clergymen, owing its life mainly to oral transmission from one generation of Huckleberry Finns to another (though in recent years the language has found literary deposition in some books designed to thrill ladies and clergymen). The 'other America' that speaks this language can be found wherever people are excluded, or exclude themselves, from the world of middle-class propriety. We find it in those sections of the working class that have not yet proceeded too far on the road of *embourgeoisement*, in slums, shanty-towns and those parts of cities that urban sociologists have called 'areas of transition'. We find it expressed powerfully in the world of the American Negro. We also come on it in the sub-worlds of those who have, for one reason or another, withdrawn voluntarily from Main Street and Madison Avenue – in the worlds of hipsters, homosexuals, hoboes and other 'marginal men', those worlds that are kept safely out of sight on the streets where the nice people live, work and amuse themselves *en famille* (though these worlds may on some occasions be rather convenient for the male of the species 'nice people' – precisely on occasions when he happily finds himself *sans famille*).

American sociology, accepted early both in academic circles and by those concerned with welfare activities, was from the beginning associated with the 'official America' with the world of policy makers in community and nation. Sociology today retains this respectable affiliation in university, business and government. The appellation hardly induces eyebrows to be raised, except the eyebrows of such Southern racists sufficiently literate to have read the footnotes of the desegregation decision of 1954. However, we would contend that there has been an important undercurrent in American sociology, relating it to that 'other America' of dirty language and disenchanted attitudes, that state of mind that refuses to be impressed, moved or be-fuddled by the official ideologies.

This unrespectable perspective on the American scene can be seen most clearly in the figure of Thorstein Veblen, one of the early important sociologists in America. His biography itself constitutes an exercise in marginality: a difficult, querulous character; born on a Norwegian farm on the Wisconsin frontier; acquiring English as a foreign language; involved all his life with morally and politically suspect individuals; an academic migrant; an inveterate seducer of other people's women. The perspective on America gained from this angle of vision can be found in the unmasking satire that runs like a purple thread through Veblen's work, most famously in his *Theory of the Leisure Class*, that merciless look from the underside at the pre-tensions of the American *haute bourgeoisie*. Veblen's view of society can be understood most easily as a series of non-Rotarian insights – his understanding of 'conspicuous con-sumption' as against the middle-class enthusiasm for the 'finer things', his analysis of economic processes in terms of manipulation and waste as against the American pro-ductivity ethos, his understanding of the machinations of real estate speculation as against the American community ideology, most bitterly his description of academic life (in *The Higher Learning in America*) in terms of fraud and flatulence as against the American cult of education. We are not associating ourselves here with a certain neo-Veblenism that has become fashionable with some younger American sociologists, nor arguing that Veblen was a giant in the development of the field. We are only pointing to his irreverent curiosity and clear-sightedness as marks of a per-spective coming from those places in the culture in which one gets up to shave about noon on Sundays. Nor are we arguing that clear-sightedness is a general trait of unrespect-ability. Stupidity and sluggishness of thought are probably distributed quite fairly throughout the social spectrum. But where there is intelligence and where it manages to free itself from the goggles of respectability, we can expect a clearer view of society than in those cases where the oratorical imagery is taken for real life.

A number of developments in empirical studies in American sociology furnish evidence of this same fascination with the unrespectable view of society. For example, looking back at the powerful development of urban studies undertaken at the University of Chicago in the 1920s we are struck by the apparently irresistible attraction to the seamier sides of city life upon these researchers. The advice to his students of Robert Park, the most important figure in this development, to the effect that they should get their hands dirty with research often enough meant quite literally an intense interest in all the things that North Shore residents would call 'dirty'. We sense in many of these studies the excitement of discovering the picaresque undersides of the metropolis – studies of slum life, of the melancholy world of rooming houses, of Skid Row, of the worlds of crime and prostitution. One of the offshoots of this so-called 'Chicago School' has been the sociological study of occupations, due very largely to the pioneering work of Everett Hughes and his students. Here also we find a fascination with every possible world in which human beings live and make a living, not only with the worlds of the respectable occupations, but with those of the taxi driver, the apartment-house janitor, the professional boxer or the jazz musician. The same tendency can be discovered in the course of American community studies following in the wake of the famous *Middletown* studies of Robert and Helen Lynd. Inevitably these studies had to by-pass the official versions of community life, to look at the social reality of the community not only from the perspective of city hall but also from that of the city jail. Such sociological procedure is *ipso facto* a refutation of the respectable presupposition that only certain views of the world are to be taken seriously.

We would not want to give an exaggerated impression of the effect of such investigations on the consciousness of sociologists. We are well aware of the elements of muck-raking and romanticism inherent in some of this. We also know that many sociologists participate as fully in the respectable *Weltanschauung* as all the other P T A members on their block. Nevertheless, we would maintain that socio-logical consciousness predisposes one towards an awareness of worlds other than that of middleclass respectability, an awareness which already carries within itself the seeds of intellectual unrespectability. In the second *Middletown* study the Lynds have given a classic analysis of the mind of middle-class America in their series of 'of course state-ments' – that is, statements that represent a consensus so strong that the answer to any question concerning them will habitually be prefaced with the words 'of course'. 'Is our economy one of free enterprise?' 'Of course!' 'Are all our important decisions arrived at through the democratic pro-cess?' 'Of course!' 'Is monogamy the natural form of marriage?' 'Of course!' The sociologist, however con-servative and conformist he may be in his private life, knows that there are serious questions to be raised about every one of these 'of course statements'. In this knowledge alone he is brought to the threshold of unrespectability.

This unrespectable motif of sociological consciousness need not imply a revolutionary attitude. We would even go further than that and express the opinion that sociological understanding is inimical to revolutionary ideologies, not because it has some sort of conservative bias, but because it sees not only through the illusions of the present *status quo* but also through the illusionary expectations concerning possible futures, such expectations being the customary spiritual nourishment of the revolutionary. This nonrevolutionary and moderating soberness of sociology we would value quite highly. More regrettable, from the viewpoint of one's values, is the fact that sociological understanding by itself does not necessarily lead to a greater tolerance with respect to the foibles of mankind. It is possible to view social reality with compassion or with cynicism, both attitudes being compatible with clear-sightedness. But whether he can bring himself to human sympathy with the phenomena he is studying or not, the sociologist will in some measure be detached from the taken-for-granted postures of his society. Unrespectability, whatever its ramifications in the emotions and the will, must remain a constant possibility in the sociologist's mind. It may be segregated from the rest of his life, overlaid by the routine mental states of everyday existence, even denied ideologically. Total respectability of thought, however, will invariably mean the death of sociology. This is one of the reasons why genuine sociology disappears promptly from the scene in totalitarian countries, as is well illustrated in the instance of Nazi Germany. By implication, sociological understanding is always potentially dangerous to the minds of policemen and other guardians of public order, since it will always tend to relativize the claim to absolute rightness upon which such minds like to rest.

Before concluding this chapter, we would look once more on this phenomenon of relativization that we have already touched upon a few times. We would now say explicitly that sociology is so much in tune with the temper of the modern era precisely because it represents the consciousness of a world in which values have been radically relativized. This relativization has become so much part of our everyday imagination that it is difficult for us to grasp fully how closed and absolutely binding the world views of other cultures have been and in some places still are. The American sociologist Daniel Lerner, in his study of the contemporary Middle East (*The Passing of Traditional Society*), has given us a vivid portrait of what 'modernity' means as an altogether new kind of consciousness in those countries. For the traditional mind one is what one is, where one is, and cannot even imagine how one could be anything different. The modern mind, by contrast, is mobile, participates vicariously in the lives of others differently located from oneself, easily imagines itself changing occupation or residence. Thus Lerner found that some of the illiterate respondents to his questionnaires could only respond with laughter to the question as to what they would do if they were in the position of their rulers and would not even consider the question as to the circumstances under which they would be willing to leave their native village. Another way of putting this would be to say that traditional societies assign definite and permanent identities to their members. In modern society identity itself is uncertain and in flux. One does not really know what is expected of one as a ruler, as a parent, as a cultivated person, or as one who is sexually normal. Typically, one then requires various experts to tell one. The book club editor tells us what culture is, the interior designer what taste we ought to have, and the psychoanalyst who we are. To live in modern society means to live at the centre of a kaleidoscope of everchanging roles.

Again, we must forgo the temptation of enlarging on this point, since it would take us rather far afield from our argument into a general discussion of the social psychology of modern existence. We would rather stress the intellectual aspect of this situation, since it is in that aspect that we would see an important dimension of sociological consciousness. The unprecedented rate of geographical and social mobility in modern society means that one becomes exposed to an unprecedented variety of ways of looking at the world. The insights into other cultures that one might gather by travel are brought into one's own living room through the mass media. Someone once defined urbane sophistication as being the capacity to remain quite unperturbed upon seeing in front of one's house a man dressed in a turban and a loincloth, a snake coiled around his neck, beating a tom-tom as he leads a leashed tiger down the street. No doubt there are degrees to such sophistication, but a measure of it is acquired by every child who watches television. No doubt also this sophistication is commonly only superficial and does not extend to any real grappling with alternate ways of life. Nevertheless, the immensely broadened possibility of travel, in person and through the imagination, implies at least potentially the awareness that one's own culture, including its basic values, is relative in space and time. Social mobility, that is, the movement from one social stratum to another, augments this relativizing effect. Where-ever industrialization occurs, a new dynamism is injected into the social system. Masses of people begin to change their social position, in groups or as individuals. And usually this change is in an 'upward' direction. With this movement an individual's biography often involves a considerable journey not only through a variety of social groups but through the intellectual universes that are, so to speak, attached to these groups. Thus the Baptist mail clerk who used to read the *Reader's Digest* becomes an Episcopalian junior executive who reads *The New Yorker*, or the faculty wife whose husband becomes department chairman may graduate from the best-seller list to Proust or Kafka.

In view of this overall fluidity of world views in modern society it should not surprise us that our age has been characterized as one of conversion. Nor should it be surprising that intellectuals especially have been prone to change their world views radically and with amazing frequency. The intellectual attraction of strongly presented, theoretically closed systems of thought such as Catholicism or Communism has been frequently commented upon.

Psychoanalysis, in all its forms, can be understood as an institutionalized mechanism of conversion, in which the individual changes not only his view of himself but of the world in general. The popularity of a multitude of new cults and creeds, presented in different degrees of intellectual refinement depending upon the educational level of their clientele, is another manifestation of this proneness to conversion of our contemporaries. It almost seems as if modern man, and especially modern educated man, is in a perpetual state of doubt about the nature of himself and of the universe in which he lives. In other words, the awareness of relativity, which probably in all ages of history has been the possession of a small group of intellectuals, today appears as a broad cultural fact reaching far down into the lower reaches of the social system.

We do not want to give the impression that this sense of relativity and the resulting proneness to change one's entire *Weltanschauung* are manifestations of intellectual or emotional immaturity. Certainly one should not take with too much seriousness some representatives of this pattern. Nevertheless, we would contend that an essentially similar pattern becomes almost a destiny in even the most serious intellectual enterprises. It is impossible to exist with full awareness in the modern world without realizing that moral, political and philosophical commitments are relative, that, in Pascal's words, what is truth on one side of the Pyrénées is error on the other. Intensive occupation with the more fully elaborated meaning systems available in our time gives one a truly frightening understanding of the way in which these systems can provide a total interpretation of reality, within which will be included an interpretation of the alternate systems and of the ways of passing from one system to another. Catholicism may have a theory of Communism, but Communism returns the compliment and will produce a theory of Catholicism. To the Catholic thinker the Communist lives in a dark world of materialist delusion about the real meaning of life. To the Communist his Catholic adversary is helplessly caught in the 'false consciousness' of a bourgeois mentality. To the pyschoanalyst both Catholic and Communist may simply be acting out on the intellectual level the unconscious impulses that really move them. And psychoanalysis may be to the Catholic an escape from the reality of sin and to the Communist an avoidance of the realities of society. This means that the individual's choice of viewpoint will determine the way in which he looks back upon his own biography. American prisoners of war 'brainwashed' by the Chinese Communists completely changed their viewpoints on social and political matters. To those that returned to America this change represented a sort of illness brought on by outward pressure, as a convalescent may look back on a delirious dream. But to their former captors this changed consciousness represents a brief glimmer of true understanding between long periods of ignorance. And to those prisoners who decided not to return, their conversion may still appear as the decisive passage from darkness to light.

Instead of speaking of conversion (a term with religiously charged connotations) we would prefer to use the more neutral term of 'alternation' to describe this phenomenon. The intellectual situation just described brings with it the possibility that an individual may alternate back and forth between logically contradictory meaning systems. Each time, the meaning system he enters provides him with an interpretation of his existence and of his world, including in this interpretation an explanation of the meaning system he has abandoned. Also, the meaning system provides him with tools to combat his own doubts. Catholic confessional discipline, Communist 'autocriticism' and the psychoanalytic techniques of coping with 'resistance' all fulfil the same purpose of preventing alternation out of the particular meaning system, allowing the individual to interpret his own doubts in terms derived from the system itself, thus keeping him within it. On lower levels of sophistication there will also be various means employed to cut off questions that might threaten the individual's allegiance to the system, means that one can see at work in the dialectical acrobatics of even such relatively unsophisticated groups as Jehovah's Witnesses or Black Muslims.

If one resists the temptation, however, to accept such dialectics, and is willing to face squarely the experience of relativity brought on by the phenomenon of alternation, then one comes into possession of yet another crucial dimension of sociological consciousness – the awareness that not only identities but ideas are relative to specific social locations. We shall see in a later chapter the considerable importance of this awareness for sociological understanding. Suffice it to say here that this relativizing motif is another of the fundamental driving forces of the sociological enterprise.

In this chapter we have tried to outline the dimensions of sociological consciousness through the analysis of three motifs – those of debunking, unrespectability and relativizing. To these three we would, finally, add a fourth one, much less far-reaching in its implications but useful in rounding out our picture – the cosmopolitan motif. Going back to very ancient times, it was in cities that there developed an openness to the world, to other ways of thinking and acting. Whether we think of Athens or Alexandria, of medieval Paris or Renaissance Florence, or of the turbulent urban centres of modern history, we can identify a certain cosmopolitan consciousness that was especially characteristic of city culture. The individual, then, who is not only urban but urbane is one who, however passionately he may be attached to his own city, roams through the whole wide world in his intellectual voyages. His mind, if not his body and his emotions, is at home wherever there are other men who think. We would submit that sociological consciousness is marked by the same kind of cosmopolitanism. This is why a narrow parochialism in its focus of interest is always a danger signal for the sociological venture (a danger signal that, unfortunately, we would hoist over quite a few sociological studies in America today). The sociological perspective is a broad, open, emancipated vista on human life. The sociologist, at his best, is a man with a taste for other lands, inwardly open to the measureless richness of human possibilities, eager for new horizons and new worlds of human meaning. It probably requires no additional elaboration to make the point that this type of man can play a particularly useful part in the course of events today.

3

DIGRESSION: ALTERNATION AND BIOGRAPHY

IN the preceding chapter we tried to show how sociological consciousness is particularly likely to arise in a cultural situation marked by what we have termed 'alternation', that is, the possibility to choose between varying and sometimes contradictory systems of meaning. Before we proceed to the main part of our argument, which will be an attempt to delineate certain key features of the sociological perspective on human existence, we would like to stop for one further moment at this phenomenon of 'alternation', going a little bit off our main course and asking what significance this phenomenon may have for the individual trying to understand his own biography. This digression may make clearer that sociological consciousness is not only an intriguing historical apparition that one may profitably study, but is also a live option for the individual seeking to order the events of his own life in some meaningful fashion.

The common-sense view would have it that we live through a certain sequence of events, some more and some less important, the sum of which is our biography. To compile a biography, then, is to record these events in chronological order or in the order of their importance. But even a purely chronological record raises the problem of just what events should be included, since obviously not everything the subject of the record ever did could be covered. In other words, even a purely chronological record forces one to raise questions concerning the relative importance of certain events. This becomes especially clear in deciding on what historians call 'periodization'. Just when in the history of Western civilization should one consider the Middle Ages to have begun? And just when in the biography of an individual can one assume that his youth has come to an end? Typically, such decisions are made on the basis of events that the historian or the biographer considers to have been 'turning points' – say, the coronation of Charlemagne, or the day on which Joe Blow decides to join the church and remain faithful to his wife. However, even the most optimistic historians and biographers (and, just as important, autobiographers) have moments of doubts as to the choice of these particular events as the truly decisive ones. Perhaps, they may say, it is not the coronation of Charlemagne but his conquest of the Saxons that should be taken as the great turning point. Or perhaps it was the point at which Joe gave up his ambition to become a writer that should mark the beginning of *his* middle age. The decision for one as against another event obviously depends on one's frame of reference.

This fact is not altogether hidden from common-sense thinking. It is taken care of by the notion that a certain maturity is required before one can really understand what one's life has been all about. The mature consciousness of oneself is then the one that has, so to speak, an epistemologically privileged position. The middle-aged Joe Blow, having accepted the fact that his wife will not get to be any prettier and that his job as assistant advertising manager will not become any more interesting, looks back on his past and decides that his earlier aspirations to possess many beautiful women or to write the definitive novel of the half-century were quite immature. Maturity is the state of mind that has settled down, come to terms with the *status quo*, given up the wilder dreams of adventure and fulfilment. It is not difficult to see that such a notion of maturity is psychologically functional in giving the individual a rationalization for having lowered his sights. Nor is it difficult to imagine how the young Joe, assuming the gift of augury, would have recoiled from his later self as from an image of defeat and desperation. In other words, we would contend that the notion of maturity really begs the question of what is important and what unimportant in one's biography. What may look like mellow maturity from one point of view may be interpreted as cowardly compromise from another. To become older, alas, is not necessarily to become wiser. And the perspective of today has no epistemological priority over the one of last year. Incidentally, it is this same recognition that makes most historians today wary of any notion of progress or evolution in human affairs. It is too easy to think that our own age is the epitome of what men have achieved so far, so that any past period can be judged on a scale of progress in terms of its closeness to or distance from the point at which we now stand. Perhaps the decisive event of man's history on this planet took place on a quiet afternoon in the year 2405 BC when an Egyptian priest woke up from his siesta and suddenly knew the final answer to the riddle of human existence – and promptly expired without telling anyone. Perhaps everything that has happened since is nothing but an inconsequential postlude. Nobody can possibly know, except perhaps the gods, and their communications appear to be regrettably ambiguous.

But to return from such metaphysical speculation to the problems of biography, it would seem, therefore, that the course of events that constitute one's life can be subjected to alternate interpretations. Nor can this be done only by the outside observer, so that after we're dead rival biographers may quarrel over the real significance of this or that thing we have done or said. We ourselves go on interpreting and reinterpreting our own life. As Henri Bergson has shown, memory itself is a reiterated act of interpretation. As we remember the past, we reconstruct it in accordance with our present ideas of what is important and what is not. This is what the psychologists call 'selective perception', except that they usually apply this concept to the present. This means that in any situation, with its near-infinite number of things that could be noticed, we notice only those things that are important for our immediate purposes. The rest we ignore. But in the present these things that we have ignored may be thrust upon our consciousness by someone who points them out to us. Unless we are literally mad we shall have to admit that they are there, although we may emphasize that we are not interested in them very much. But the things in the past that we have decided to ignore are much more helpless against our annihilating non-remembrance. They are not here to be pointed out to us against our will, and only in rare instances (as, for example, in criminal proceedings) are we confronted with evidence that we cannot dispute. This means that common sense is quite wrong in thinking that the past is fixed, immutable, invariable, as against the everchanging flux of the present. On the contrary, at least within our own consciousness, the past is malleable and flexible, constantly changing as our recollection reinterprets and re-explains what has happened. Thus we have as many lives as we have points of views. We keep reinterpreting our biography very much as the Stalinists kept rewriting the Soviet Encyclopedia, calling forth some events into decisive importance as others were banished to ignominious oblivion.

We can safely assume that this process of reshaping the past (which probably is inherent in the very fact of language itself) is as old as *homo sapiens*, if not his hominoid ancestors, and that it helped to while away the long millennia in which men did little but dully bang away with their fist-axes. Every rite of passage is an act of historical interpretation and every wise old man is a theorist of historical development. But what is distinctively modern is the frequency and rapidity with which such reinterpretation often occurs in the lives of many individuals, and the increasingly common situation in which different systems of interpretation can be chosen in this game of re-creating the world. As we have already pointed out in the previous chapter, the great intensification of both geographical and social mobility is a major cause of this. A few examples might serve to explicate this point further.

People on the move physically are frequently people who are also on the move in their self-understanding. Take the amazing transformations of identity and self-image that can be the result of a simple change of residence. Certain places have served as the classic locations in which such transformations are produced almost as on an assembly line. One cannot, for example, understand Greenwich Village without understanding Kansas City. Since its inception as a gathering place of those interested in changing their identity, it has served as a socio-psychological apparatus through which men and women pass as through a magical retort, going in as nice Midwesterners and coming out as nasty deviants. What was proper before is improper after, and vice versa. What used to be taboo becomes *de rigueur*, what used to be obvious becomes laughable, and what used to be one's world becomes that which must be overcome. Obviously going through such a transformation involves a reinterpretation of one's past, and a radical one at that. One now realizes that the great emotional upheavals of the past were but puerile titillations, that those whom one thought important people in one's life were but limited provincials all along. The events of which one used to be proud are now embarrassing episodes in one's prehistory. They may even be repressed from memory if they are too much at variance with the way in which one wants to think of oneself now. Thus the glowing day when one was class valedictorian makes room in one's reconstructed biography for a then unimportant-seeming evening when one first tried to paint, and instead of reckoning an era from the date when one accepted Jesus at a church summer camp one does so from that other date, previously one of anxious shame but now one of decisive self-legitimation, when one lost one's virginity in the back of a parked automobile. We go through life refashioning our calendar of holy days, raising up and tearing down again the signposts that mark our progress through time toward ever newly defined fulfilments. For it will be clear by now that no magic is so strong that it may not be overcome by a newer brand. Greenwich Village may later become but another phase in one's life, another experiment, another mistake even. Old markers may be retrieved from the debris of discarded chronologies. For example, the conversion experience in the church camp may later turn out to have been the first uncertain groping towards the truth one has now realized fully in becoming a Catholic. And completely new ordering categories may be imposed on the same past. Thus, for example, one may discover in one's psychoanalysis that both conversion and sexual initiation, both those things of which one was proud and those of which one was ashamed, and both one's earlier and one's later interpretations of these events, were part and parcel of the same neurotic syndrome. And so on *ad infinitum* – and *ad nauseam*.

To avoid giving the preceding paragraphs the appearance of a Victorian novel we have been miserly in the use of quotation marks. All the same, it should be clear now that it was with tongue in cheek that we spoke of this being 'realized' or that being 'discovered'. The 'true' understanding of our past is a matter of our viewpoint. And, obviously, our viewpoint may change. 'Truth', then, is not only a matter of geography but of the time of day. Today's 'insight' becomes tomorrow's 'rationalization', and the other way around.

Social mobility (the movement from one level of society to another) has very similar consequences in terms of the reinterpretation of one's life as geographical mobility. Take

73

the way in which a man's self-image changes as he moves up the social ladder. Perhaps the saddest aspect of this change is the way in which he now reinterprets his relationships to the people and events that used to be closest to him. For example, everything connected with the Little Italy of one's childhood undergoes a malevolent mutation when viewed from the vantage point of the suburban home that one has finally clawed one's way to. The girl of one's teenage dreams is transmuted into an ignorant though pretty peasant. Boyhood friendships become irritating reminders of an embarrassing former self, long left behind, together with old ideas of honour, magic and street-corner patriotism. Even Mamma, who used to be the orb around which the universe revolved, has become a silly old Italian woman one must pacify occasionally with the fraudulent display of an old self that no longer exists. Again, there are elements in this picture that are probably as old as mankind, since presumably the end of childhood has always meant an eclipse of gods. What is new is that so many children in our kind of society not only grow into adulthood but, in doing so, move into social worlds utterly beyond the comprehension of their parents. This is an inevitable consequence of massive social mobility. American society having been one of high mobility for quite some time, many Americans seemingly spend years of their life reinterpreting their own background, retelling over and over again (to themselves *and* to others) the story of what they have been and what they have become – and in the process killing their parents in a sacrificial ritual of the mind. Needless to add, the phrases 'what they have been' and 'what they have become' belong in quotation marks! It is no wonder, incidentally, that the Freudian mythology of parricide has found ready credence in American society and especially in those recently middle-class segments of it to whom such rewriting of biographies is a social necessity of legitimating one's hard-won status.

Instances of geographical and social mobility merely illustrate more sharply a process that goes on throughout society and in many different social situations. The confessing husband who reinterprets the love affairs of his past to bring them into a line of ascent culminating in his marriage, the newly divorced wife who reinterprets her marriage *ab initio* in such a way that each stage of it serves to explain the final fiasco, the inveterate gossiper who reinterprets his various relationships in each new gossiping group he enters (explaining his relationship to A in a certain way to B, making it appear that B is his real intimate, then turning around and sacrificing this supposed intimacy by gossiping about B to A, and so on), the man who has discovered deceit in one he trusted and now pretends that he had always been suspicious of him (pretending this to himself as much as to others) – all these are engaged in the same perennial pastime of correcting fortune by remaking history. Now, in most of these cases, the process of reinterpretation is partial and at best half-conscious. One rectifies the past where one has to, leaving untouched what one can incorporate into one's present self-image. And these continuous modifications and adjustments in one's biographical *tableau* are rarely integrated in a clear, consistent definition of oneself. Most of us do not set out deliberately to paint a grand portrait of ourselves. Rather we stumble like drunkards over the sprawling canvas of our self-conception, throwing a little paint here, erasing some lines there, never really stopping to obtain a view of the likeness we have produced. In other words, we might accept the existentialist notion that we create ourselves if we add the observation that most of this creation occurs haphazardly and at best in half-awareness.

There are some cases, however, where the reinterpretation of the past is part of a deliberate, fully conscious and intellectually integrated activity. This happens when the reinterpretation of one's biography is one aspect of conversion to a new religious or ideological *Weltanschauung,* that is, a universal meaning system *within which* one's biography can be located. Thus the convert to a religious faith can now understand his entire previous life as a providential movement towards the moment when the mist lifted from before his eyes. Classic statements of this would be Augustine's *Confessions* or Newman's *Apologia Pro Vita Sua.* Conversion introduces a new periodization in one's biography – BC and AD, pre-Christian and Christian, pre-Catholic and Catholic. Inevitably the period coming before the event now designated as decisive is interpreted as a preparation. The prophets of the old dispensation are designated as forerunners and forecasters of the new. In other words, conversion is an act in which *the past* is dramatically transformed.

Satori, the experience of illumination sought in Zen Buddhism, is described as 'seeing things with new eyes'. While this is manifestly apt with regard to religious conversions and mystic metamorphoses, the modern secular faiths provide very similar experiences for their adherents. The process of becoming a Communist, for instance, involves a drastic reassessment of one's past life. Just as the new Christian now understands his previous life as a long night of sin and alienation from the saving truth, so the young Communist understands his past as a captivity in the 'false consciousness' of a bourgeois mentality. Past events must be reinterpreted radically. What used to be carefree joy is now classified under the sin of pride, or what used to be personal integrity is now seen as bourgeois sentimentality. Consequently, past relationships must be reappraised too. Even the love of one's parents may have to be discarded as a temptation to apostasy or as treason to the party.

Psychoanalysis provides for many people in our society a similar method of ordering the discrepant fragments of their biography in a meaningful scheme. This method is particularly functional in a comfortable middle-class society, too 'mature' for the courageous commitment demanded by religion or revolution. Containing within its system an elaborate and supposedly scientific means of explaining all human behaviour, psychoanalysis gives its adherents the luxury of a convincing picture of themselves without making any moral demands on them and without upsetting their socio-economic applecarts. This is evidently a technological improvement in conversion management as compared with Christianity or Communism. Apart from that, the reinterpretation of the past proceeds in analogous fashion. Fathers, mothers, brothers, sisters, wives and children are thrown one by one into the conceptual cauldron and emerge as metamorphosed figures of the Freudian pantheon. Oedipus takes Jocasta to the movies and beholds the Primal Father over the breakfast table. And, once more, everything makes sense now.

The experience of conversion to a meaning system that is capable of ordering the scattered data of one's biography is liberating and profoundly satisfying. Perhaps this has its roots in a deep human need for order, purpose and intelligibility. However, the dawning recognition that this or any other conversion is not necessarily final, that one could be reconverted and re-reconverted, is one of the most terrifying ideas the mind can have. The experience of what we have called 'alternation' (which is precisely the perception of oneself in front of an infinite series of mirrors, each one transforming one's image in a different potential conversion) leads to a feeling of vertigo, a metaphysical agoraphobia before the endlessly overlapping horizons of one's possible being. It would be most gratifying if we could now produce sociology as the magic pill that can be swallowed so that all these horizons promptly fall into place. If we did that, we would simply be adding one more mythology to all the others that promise relief from the epistemological anxieties of the 'alternation' sickness. The sociologist, *qua* sociologist, cannot offer any such salvation (he may be a *guru* in his extra-curricular activities, but that need not concern us here). He is just like any other man in that he must exist in a situation where information about the ultimate meaning of things is sparse, often clearly spurious and probably never overwhelming. He has no epistemological miracles for sale. Indeed, the sociological frame of reference is but another system of interpretation that can be applied to existence and that can be superseded again in other attempts at biographical hermeneutics.

Nevertheless, the sociologist can provide a very simple and therefore all the more useful insight to men trying to find their way through the jungle of competing world views. That is the insight that every one of these world views is *socially grounded*. To put this a little differently, every *Weltanschauung* is a conspiracy. The conspirators are those who construct a social situation in which the particular world view is taken for granted. The individual who finds himself in this situation becomes more prone every day to share its basic assumptions. That is, we change our world views (and thus our interpretations and reinterpretations of our biography) as we move from one social world to another. Only the madman or the rare case of genius can inhabit a world of meaning all by himself. Most of us acquire our meanings from other men and require their constant support so that these meanings may continue to be believable. Churches are agencies for the mutual reinforcement of meaningful interpretations. The beatnik must have a beatnik sub-culture, as must the pacifist, the vegetarian and the Christian Scientist. But the fully adjusted, mature, middle-of-the-road, sane and sensible suburbanite also requires a specific social context that will approve and sustain his way of life. Indeed, every one of these terms –

'adjustment', 'maturity', 'sanity', and so on – refers to socially relative situations and becomes meaningless when divorced from these. One adjusts to a particular society. One matures by becoming habituated to it. One is sane if one shares its cognitive and normative assumptions.

Individuals who change their meaning systems must, therefore, change their social relationships. The man who redefines himself by marrying a certain woman must drop the friends that do not fit this self-definition. The Catholic marries a non-Catholic at the peril of his Catholicism, just as the beatnik endangers *his* ideology by having lunch too often with his uptown agent. Meaning systems are socially constructed. The Chinese 'brainwasher' conspires with his victim in fabricating a new life-story for the latter, just as does the psychoanalyst with his patient. Of course, in both situations the victim/patient comes to believe that he is 'discovering' truths about himself that were there long before this particular conspiracy got under way. The sociologist will be, at the very least, sceptical about this conviction. He will strongly suspect that what appears as discovery is really invention. And he will know that the plausibility of what is thus invented is in direct relation to the strength of the social situation within which the invention in concocted.

In a later chapter we shall enlarge further on this vexing connexion between what we think and who we sup with. In this digression we have merely tried to show that the experience of relativity and 'alternation' is not only a global historical phenomenon but a real existential problem in the life of the individual. The insight of sociology into the social roots of this experience may be slight comfort to those who would find a philosophical or theological answer to the agonizing problem thus posed. But in this world of painfully rationed revelations one ought to be grateful for small favours. The sociological perspective, with its irritating interjection of the question *Says who?* into the grand debate of *Weltanschauungen,* introduces an element of sober scepticism that has an immediate utility in giving some protection at least against converting too readily. Sociological consciousness moves in a frame of reference that allows one to perceive one's biography as a movement within and through specific social worlds, to which specific meaning systems are attached. This by no means solves the problem of truth. But it makes us a little less likely to be trapped by every missionary band we encounter on the way.

4

MAN IN SOCIETY

At a certain age children are greatly intrigued by the possibility of locating themselves on a map. It appears strange that one's familiar life should actually have all occurred in an area delineated by a set of quite impersonal (and hitherto unfamiliar) coordinates on the surface of the map. The child's exclamations of 'I was there' and 'I am here right now' betray the astonishment that the place of last summer's vacation, a place marked in memory by such sharply personal events as the ownership of one's first dog or the secret assemblying of a collection of worms, should have specific latitudes and longitudes devised by strangers to one's dog, one's worms, and oneself. This locating of oneself in configurations conceived by strangers is one of the important aspects of what, perhaps euphemistically, is called 'growing up'. One participates in the real world of grown-ups by having an address. The child who only recently might have mailed a letter addressed 'To my Granddaddy' now informs a fellow worm-collector of his exact address – street, town, state and all – and finds his tentative allegiance to the grown-up world view dramatically legitimated by the arrival of the letter.

As the child continues to accept the reality of this world view, he continues to gather addresses – 'I'm six years old', 'My name is Brown, like my father's, that's because my parents are divorced', 'I'm a Presbyterian', 'I'm an American', and eventually perhaps 'I'm in the special class for the bright kids, that's because my IQ is 130'. The horizons of the world, as the grown-ups define it, are determined by the coordinates of remote map-makers. The child may produce alternate identifications by having himself addressed as father while playing house, as Indian chief or as Davy Crockett, but he will know all the time that he is only playing and that the real facts about himself are those registered by the school authorities. We leave out the quotation marks and thus betray that we too have been trapped into sanity in our childhood – of course, we would write all the key words in quotation marks – 'know', 'real', 'facts'. The sane child is the one who believes in what it says in the school records. The normal adult is the one who lives within his assigned coordinates.

What is called the common-sense view is actually the grown-up view taken for granted. It is a matter of the school records having become an ontology. One now identifies one's being as a matter of course with the way one is pinpointed on the social map. What this means for one's identity and one's ideas will be pursued further in the next chapter. What interests us at the moment is the way in which such location tells an individual just what he may do and what he can expect of life. To be located in society means to be at the intersection point of specific social forces. Commonly one ignores these forces at one's peril. One moves within society within carefully defined systems of power and prestige. And once one knows how to locate oneself, one also knows that there is not an awful lot that one can do about this.

The way in which lower-class individuals use the pronouns 'they' and 'them' nicely expresses this consciousness of the heteronomy of one's life. 'They' have things fixed in a certain way, 'they' call the tune, 'they' make the rules. This concept of 'them' may not be too easily identified with particular individuals or groups. It is 'the system', the map made by strangers, over which one must keep crawling. But it would be a onesided way of looking at 'the system' if one assumes that this concept loses its meaning as one moves into the higher levels of society. To be sure, there will be a greater sense of freedom of movement and decision, and realistically so. But the basic co-ordinates within which one can move and decide have still been drawn by others, most of them strangers, many of them long in their graves. Even the total autocrat exercises his tyranny against constant resistance, not necessarily political resistance, but the resistance of custom, convention and sheer habit. Institutions carry within them a principle of inertia, perhaps founded ultimately on the hard rock of human stupidity. The tyrant finds that even if nobody dares act against him, his orders will still be nullified again and again by simple lack of comprehension. The alien-made fabric of society reasserts itself even against terror. But let us leave the question of tyranny. On the levels occupied by most men, including the writer and (we daresay) almost all the readers of these lines, location in society constitutes a definition of rules that have to be obeyed.

As we have seen, the common-sense view of society understands this. The sociologist does not contradict this understanding. He sharpens it, analyzes its roots, sometimes either modifies or extends it. We shall see later that sociological perspective finally goes beyond the common-sense understanding of 'the system' and our captivity in it. But in most specific social situations that the sociologist sets out to analyze he will find little reason to quarrel with the notion that 'they' are in charge. On the contrary, 'they' will loom larger and in more pervasive fashion over our lives than we thought before the sociological analysis. This aspect of sociological perspective can be clarified by looking at two important areas of investigation – social control and social stratification.

Social control is one of the most generally used concepts in sociology. It refers to the various means used by a society to bring its recalcitrant members back into line. No society can exist without social control. Even a small group of people meeting but occasionally will have to develop their mechanisms of control if the group is not to dissolve in a very short time. It goes without saying that the instrumentalities of social control vary greatly from one social situation to another. Opposition to the line in a business organization may mean what personnel directors call a terminal interview, and in a criminal syndicate a terminal automobile ride. Methods of control vary with the purpose and character of the group in question. In either case, control mechanisms function to eliminate undesirable personnel and (as it was put classically by King Christophe of Haiti when he had every tenth man in his forced-labour battalion executed) 'to encourage the others'.

The ultimate and, no doubt, the oldest means of social control is physical violence. In the savage society of children it is still the major one. But even in the politely operated societies of modern democracies the ultimate argument is violence. No state can exist without a police force or its equivalent in armed might. This ultimate violence may not be used frequently. There may be innumerable steps before its application, in the way of warnings and reprimands. But if all the warnings are disregarded, even in so slight a matter as paying a traffic ticket, the last thing that will happen is that a couple of cops show up at the door with handcuffs and a Black Maria. Even the moderately courteous cop who hands out the initial traffic ticket is likely to wear a gun – just in case. And even in England, where he does not in the normal course of events, he will be issued one if the need arises.

In Western democracies, with their ideological emphasis on voluntary compliance with popularly legislated rules, this constant presence of official violence is under-emphasized. It is all the more important to be aware of it. Violence is the ultimate foundation of any political order. The common-sense view of society senses this, and this may have something to do with the widespread popular reluctance to eliminate capital punishment from the criminal law (though this reluctance is probably based in equal measure on stupidity, superstition and the congenital bestiality that jurists share with the bulk of their fellow citizens). However, the statement that political order rests ultimately on violence is just as true in states that have abolished capital punishment. Under certain circumstances the use of their weapons is permitted to state troopers in Connecticut, where (much to their freely expressed gratification) an electric chair graces the central penal establishment, but the same possibility exists for their colleagues in Rhode Island, where police and prison authorities have to get along without this facility. It goes without saying that in countries with less of a democratic and humanitarian ideology the instruments of violence are much less gingerly displayed – and employed.

Since the constant use of violence would be impractical and also ineffective, the official organs of social control rely mostly on the restraining influence of the generally known availability of the means of violence. For various reasons this reliance is usually justified in any society that is not on the brink of catastrophic dissolution (as, say, in situations of revolution, military defeat or natural disaster). The most important reason for this is the fact that, even in dictatorial and terroristic states, a régime tends to gain acceptance and even acceptability by the simple passage of time. This is not the place to go into the socio-psychological dynamics of this fact. In democratic societies there is at least the tendency for most people to share the values on behalf of which the means of violence are employed (this does not mean that these values have to be fine – the majority of the white people in some Southern communities may be, for instance, in favour of using violence, as administered by the police agencies, in order to uphold segregation – but it does mean that the employment of the means of violence is approved by the bulk of the populace). In any functioning society violence is used economically and as a last resort, with the mere threat of this ultimate violence sufficing for the day-to-day exercise of social control. For our purposes in this argument, the most important matter to underline is that nearly all men live in social situations in which, if all other means of coercion fail, violence may be officially and legally used against them.

If the role of violence in social control is thus understood, it becomes clear that the, so to speak, penultimate means of coercion are more important for more people most of the time. While there is a certain uninspired sameness about the methods of intimidation thought up by jurists and policemen, the less-than-violent instrumentalities of social control show great variety and sometimes imagination. Next in line after the political and legal controls one should probably place economic pressure. Few means of coercion are as effective as those that threaten one's livelihood or profit. Both management and labour effectively use this threat as an instrumentality of control in our society. But economic means of control are just as effective outside the institutions properly called the economy. Universities or churches use economic sanctions just as effectively in restraining their personnel from engaging in deviant behaviour deemed by the respective authorities to go beyond the limits of the acceptable. It may not be actually illegal for a minister to seduce his organist, but the threat of being barred forever from the exercise of his profession will be a much more effective control over this temptation than the possible threat of going to jail. It is undoubtedly not illegal for a minister to speak his mind on issues that the ecclesiastical bureaucracy would rather have buried in silence, but the chance of spending the rest of his life in minimally paid rural parishes is a very powerful argument indeed.

Naturally such arguments are employed more openly in economic institutions proper, but the administration of economic sanctions in churches or universities is not very different in its end results from that used in the business world.

Where human beings live or work in compact groups, in which they are personally known and to which they are tied by feelings of personal loyalty (the kind that sociologists call primary groups), very potent and simultaneously very subtle mechanisms of control are constantly brought to bear upon the actual or potential deviant. These are the mechanisms of persuasion, ridicule, gossip and opprobrium. It has been discovered that in group discussions going on over a period of time individuals modify their originally held opinions to conform to the group norm, which corresponds to a kind of arithmetic mean of all the opinions represented in the group. Where this norm lies obviously depends on the constituency of the group. For example, if you have a group of twenty cannibals arguing over cannibalism with one non-cannibal, the chances are that in the end he will come to see their point and, with just a few face-saving reservations (concerning, say, the consumption of close relatives), will go over completely to the majority's point of view. But if you have a group discussion between ten cannibals who regard human flesh aged over sixty years as too tough for a cultivated palate and ten other cannibals who fastidiously draw the line at fifty, the chances are that the group will eventually agree on fifty-five as the age that divides the *déjeuncr* from the *débris* when it comes to sorting out prisoners. Such are the wonders of group dynamics. What lies at the bottom of this apparently inevitable pressure towards consensus is probably a profound human desire to be accepted, presumably by whatever group is around to do the accepting. This desire can be manipulated most effectively, as is well known by group therapists, demagogues and other specialists in the field of consensus engineering.

Ridicule and gossip are potent instruments of social control in primary groups of all sorts. Many societies use ridicule as one of the main controls over children – the child conforms not for fear of punishment but in order not to be laughed at. Within our own larger culture, 'kidding' in this way has been an important disciplinary measure among Southern Negroes. But most men have experienced the freezing fear of making oneself ridiculous in some social situation. Gossip, as hardly needs elaboration, is especially effective in small communities, where most people live their lives in a high degree of social visibility and inspectability by their neighbours. In such communities gossip is one of the principal channels of communication, essential for the maintenance of the social fabric. Both ridicule and gossip can be manipulated deliberately by any intelligent person with access to their lines of transmission.

Finally, one of the most devastating means of punishment at the disposal of a human community is to subject one of its members to systematic opprobrium and ostracism. It is somewhat ironic to reflect that this is a favourite control mechanism with groups opposed on principle to the use of violence. An example of this would be 'shunning' among the Amish Mennonites. An individual who breaks one of the principal taboos of the group (for example, by getting sexually involved with an outsider) is 'shunned'. This means that, while permitted to continue to work and live in the community, not a single person will speak to him ever. It is hard to imagine a more cruel punishment. But such are the wonders of pacifism.

One aspect of social control that ought to be stressed is the fact that it is frequently based on fraudulent claims. Later, we shall take up further the general importance of fraud in a sociological understanding of human life; here we will simply stress that a conception of social control is incomplete and thus misleading unless this element is taken into account. A little boy can exercise considerable control over his peer group by having a big brother who, if need be, can be called upon to beat up any opponents. In the absence of such a brother, however, it is possible to invent one. It will then be a question of the public-relations talents of the little boy as to whether he will succeed in translating his invention into actual control. In any case, this is definitely possible. The same possibilities of fraudulence are present in all the forms of social control discussed. This is why intelligence has some survival value in the competition with brutality, malice and material resources. We shall return to this point later.

It is possible, then, to perceive oneself as standing at the centre (that is, at the point of maximum pressure) of a set of concentric circles, each representing a system of social control. The outer ring might well represent the legal and political system under which one is obligated to live. This is the system that, quite against one's will, will tax one, draft one into the military, make one obey its innumerable rules and regulations, if need be put one in prison, and in the last resort will kill one. One does not have to be a right-wing Republican to be perturbed by the ever-increasing expansion of this system's power into every conceivable aspect of one's life. A salutary exercise would be to note down for the span of a single week all the occasions, including fiscal ones, in which one came up against the demands of the politico-legal system. The exercise can be concluded by adding up the sum total of fines and/or terms of imprisonment that disobedience to the system might lead to. The consolation, incidentally, with which one might recover from this exercise would consist of the recollection that law-enforcement agencies are normally corrupt and of only limited efficiency.

Another system of social control that exerts its pressures towards the solitary figure in the centre is that of morality. custom and manners. Only the most urgent-seeming (to the authorities, that is) aspects of this system are endowed with legal sanctions. This does not mean, however, that one can safely be immoral, eccentric or unmannered. At this point all the other instrumentalities of social control go into action. Immorality is punished by loss of one's job, eccentricity by the loss of one's chances of finding a new one, bad manners by remaining uninvited and uninvitable in the groups that respect what they consider good manners. Unemployment and loneliness may be minor penalties compared to being dragged away by the cops, but they may not actually appear so to the individuals thus punished. Extreme defiance against the *mores* of our particular society, which is quite sophisticated in its control apparatus, may lead to yet another consequence – that of being defined, by common consent, as 'sick'.

Enlightened bureaucratic management (such as, for example, the ecclesiastical authorities of some Protestant denominations) no longer throws its deviant employees out on the street, but instead compels them to undergo treatment by its consulting psychiatrists. In this way, the deviant individual (that is, the one who does not meet the criteria of normality set up by management, or by his bishop) is still threatened with unemployment and with the loss of his social ties, but in addition he is also stigmatized as one who might very well fall outside the pale of responsible men altogether, unless he can give evidence of remorse ('insight') and resignation ('response to treatment'). Thus the innumerable 'counselling', 'guidance', and 'therapy' programmes developed in many sectors of contemporary institutional life greatly strengthen the control apparatus of the society as a whole and especially those parts of it where the sanctions of the politico-legal system cannot be invoked.

But in addition to these broad coercive systems that every individual shares with vast numbers of fellow controllees, there are other and less extensive circles of control to which he is subjected. His choice of an occupation (or, often more accurately, the occupation in which he happens to end up) inevitably subordinates the individual to a variety of controls, often stringent ones. There are the formal controls of licensing boards, professional organizations and trade unions – in addition, of course, to the formal requirements set by his particular employers. Equally important are the informal controls imposed by colleagues and co-workers. Again, it is hardly necessary to elaborate overly on this point. The reader can construct his own examples – the physician who participates in a prepaid comprehensive health insurance programme, the undertaker who advertises inexpensive funerals, the engineer in industry who does not allow for planned obsolescence in his calculations, the minister who says that he is not interested in the size of the membership of his church (or rather, the one who acts accordingly – they nearly all say so), the government bureaucrat who consistently spends less than his allotted budget, the assembly-line worker who exceeds the norms regarded as acceptable by his colleagues, and so on. Economic sanctions are, of course, the most frequent and effective ones in these instances – the physician finds himself barred from all available hospitals, the undertaker may be expelled from his professional organization for 'unethical conduct', the engineer may have to volunteer for the Peace Corps, as may the minister and the bureaucrat (in, say, New Guinea, where there is as yet no planned obsolescence, where Christians are few and far between, and where the governmental machinery is small enough to be relatively rational), and the assembly-line worker may find that all the defective parts of machinery in the entire plant have a way of congregating on his workbench. But the sanctions of social exclusion, contempt and ridicule may be almost as hard to bear. Each occupational role in society, even in very humble jobs, carries with it a code of conduct that is very hard indeed to defy. Adherence to this code is normally just as essential for one's career in the occupation as technical competence or training.

The social control of one's occupational system is so important because the job decides what one may do in most of the rest of one's life – which voluntary associations one will be allowed to join, who will be one's friends, where one will be able to live. However, quite apart from the pressures of one's occupation, one's other social involvements also entail control systems, many of them less unbending than the occupational one, but some even more so. The codes governing admission to and continued membership in many clubs and fraternal organizations are just as stringent as those that decide who can become an executive at IBM (sometimes, luckily for the harassed candidate, the requirements may actually be the same). In less exclusive associations, the rules may be more lax and one may only rarely get thrown out, but life can be so thoroughly unpleasant for the persistent nonconformist to the local folk-ways that continued participation becomes humanly impossible. The items covered by such unwritten codes will, naturally, vary greatly. They may include ways of dressing, language, aesthetic taste, political or religious convictions, or simply table manners. In all these cases, however, they constitute control circles that effectively circumscribe the range of the individual's possible actions in the particular situation.

Finally, the human group in which one's so-called private life occurs, that is the circle of one's family and personal friends, also constitutes a control system. It would be a grave error to assume that this is necessarily the weakest of them all just because it does not possess the formal means of coercion of some of the other control systems. It is in

this circle that an individual normally has his most important social ties. Disapproval, loss of prestige, ridicule or contempt in this intimate group has far more serious psychological weight than the same reactions encountered elsewhere. It may be economically disastrous if one's boss finally
concludes that one is a worthless nobody, but the psychological effect of such a judgment is incomparably more
devastating if one discovers that one's wife has arrived at
the same conclusion. What is more, the pressures of this
most intimate control system can be applied at those times
when one is least prepared for them. At one's job one is
usually in a better position to brace oneself, to be on one's
guard and to pretend than one is at home. Contemporary
American 'familism', a set of values that strongly emphasizes the home as a place of refuge from the tensions of the
world and of personal fulfilment, contributes effectively to
this control system. The man who is at least relatively prepared psychologically to give battle in his office is willing
to do almost anything to preserve the precarious harmony
of his family life. Last but not least, the social control of
what German sociologists have called the 'sphere of the
intimate' is particularly powerful because of the very factors
that have gone into its construction in the individual's
biography. A man chooses a wife and a good friend in acts
of essential self-definition. His most intimate relationships
are those that he must count upon to sustain the most
important elements of his self-image. To risk, therefore, the
disintegration of these relationships means to risk losing
himself in a total way. It is no wonder then that many an
office despot promptly obeys his wife and cringes before the
raised eyebrows of his friends.

If we return once more to the picture of an individual
located at the centre of a set of concentric circles, each one
representing a system of social control, we can understand
a little better that location in society means to locate one-

self with regard to many forces that constrain and coerce one. The individual who, thinking consecutively of all the people he is in a position to have to please, from the Collector of Internal Revenue to his mother-in-law, gets the idea that all of society sits right on top of him, had better not dismiss that idea as a momentary neurotic derangement. The sociologist, at any rate, is likely to strengthen him in this conception, no matter what other counsellors may tell him to snap out of it.

Another important area of sociological analysis that may serve to explicate the full meaning of location in society is that of social stratification. The concept of stratification refers to the fact that any society will consist of levels that relate to each other in terms of superordination and subordination, be it in power, privilege or prestige. To say this more simply, stratification means that every society has a system of ranking. Some strata rank higher, some lower. Their sum constitutes the stratification system of that particular society.

Stratification theory is one of the most complex sectors of sociological thought, and it would be quite outside the present context to give any kind of introduction to it. Suffice it to say that societies differ greatly in the criteria by which individuals are assigned to the different levels and that distinct stratification systems, using quite different criteria of placement, may coexist in the same society. Clearly very different factors decide an individual's position in the stratification scheme of traditional Hindu caste society from those that determine his position in a modern Western society. And the three principal rewards of social position – power, privilege and prestige – frequently do not overlap but exist side by side in distinct stratification systems. In America, wealth often leads to political power, but not inevitably so. Also, there are powerful individuals with little wealth. And prestige may be connected with activities quite unrelated to economic or political rank. These remarks may serve a cautionary purpose as we proceed to look at the way in which location in society involves the stratification system, with its enormous influence on one's entire life.

The most important type of stratification in contemporary Western society is the class system. The concept of class, as most concepts in stratification theory, has been defined in different ways. For our purposes it is sufficient to understand class as a type of stratification in which one's general position in society is basically determined by economic criteria. In such a society the rank one achieves is typically more important than the one into which one was born (although most people recognize that the latter greatly influences the former). Also, a class society is one in which there is typically a high degree of social mobility. This means that social positions are not immutably fixed, that many people change their positions for better or for worse in the course of their lifetime, and that, consequently, no position seems quite secure. As a result, the symbolic accoutrements of one's position are of great importance. That is, by the use of various symbols (such as material objects, styles of demeanour, taste, and speech, types of association and even appropriate opinions) one keeps on showing to the world just where one has arrived. This is what sociologists call status symbolism, and it has been an important concern in stratification studies.

Max Weber has defined class in terms of the expectations in life that an individual may reasonably have. In other words, one's class position yields certain probabilities, or life chances, as to the fate one may expect in society. Everyone recognizes that this is so in strictly economic terms. An upper-middle-class American of, say, twenty-five years of age has a much better chance of owning a suburban home, two cars and a cottage on the Cape ten years hence than his contemporary occupying a lower-middle-class position. This does not mean that the latter has no chance at all of achieving these things, but simply that he is operating under a statistical handicap. This is hardly surprising since class was defined in economic terms to begin with and the normal economic process ensures that those who have will have more added thereunto. But class determines life chances in ways that go far beyond the economic in its proper sense. One's class position determines the amount of education one's children are likely to receive. It determines the standards of medical care enjoyed by oneself and one's family, and, therefore, one's life expectancy – life chances in the literal sense of the word. The higher classes in our society are better fed, better housed, better educated, and live longer than their less fortunate fellow citizens. These observations may be truisms, but they gain in impact if one sees that there is a statistical correlation between the quantity of money one earns *per annum* and the number of years one may expect to do so on this earth. But the import of location within the class system goes even farther than that.

Different classes in our society not only live differently quantitatively, they live in different styles qualitatively. A sociologist worth his salt, if given two basic indices of class such as income and occupation, can make a long list of predictions about the individual in question even if no further information has been given. Like all sociological predictions, these will be statistical in character. That is, they will be probability statements and will have a margin of error. Nevertheless, they can be made with a good deal of assurance. Given these two items of information about a particular individual, the sociologist will be able to make intelligent guesses about the part of town in which the individual lives, as well as about the size and style of his house. He will also be able to give a general description of the interior decorating of the house and make a guess about the types of pictures on the wall and books or magazines likely to be found on the shelves of the living room. Moreover, he will be able to guess what kind of music the individual in question likes to listen to, and whether he does so at concerts, on a phonograph, or on the radio. But the sociologist can go on. He can predict which voluntary associations the individual has joined and where he has his church membership. He can estimate the individual's vocabulary, lay down some rough rules for his syntax and other uses of language. He can guess the individual's political affiliation and his views on a number of public issues. He can predict the number of children sired by his subject and also whether the latter has sexual relations with his wife with the lights on or off. He will be able to make some statements about the likelihood that his subject will come down with a number of diseases, physical as well as mental. As we have seen already, he will be able to place the man on an actuary's table of life expectancies. Finally, if the sociologist should decide to verify all these guesses and ask the individual in question for an interview, he can estimate the chance that the interview will be refused.

Many of the elements just touched upon are enforced by external controls in any given class milieu. Thus the corporation executive who has the 'wrong' address and the 'wrong' wife will be subjected to considerable pressures to change both. The working-class individual who wants to join an upper-middle-class church will be made to understand in unmistakable terms that he 'would be happier elsewhere'. Or the lower-middle-class child with a taste for chamber music will come up against strong pressures to change this aberration to musical interests more in accord with those of his family and friends. However, in many of these cases the application of external controls is quite unnecessary because the likelihood that deviance will occur is very small indeed. Most individuals to whom an executive career is open marry the 'right' kind of wife (the one that David Riesman has called the 'station-wagon type') almost by instinct, and most lower-middle-class children have their musical tastes formed early in such a way that they are relatively immune to the blandishments of chamber music. Each class milieu forms the personality of its constituency by innumerable influences beginning at birth and leading up to graduation from prep school or reformatory, as the case may be. Only when these formative influences have somehow failed to achieve their objective is it necessary for the mechanisms of social control to go into action. In trying to understand the weight of class, then, we are not only looking at another aspect of social control but are beginning to catch a glimpse of the way in which society penetrates the insides of our consciousness, something that we shall discuss further in the next chapter.

It should be emphasized at this point that these observations on class are in no way intended as an indignant indictment of our society. There are certainly some aspects of class differences that could be modified by certain kinds of social engineering, such as class discrimination in education and class inequalities in medical care. But no amount of social engineering will change the basic fact that different social milieus exert different pressures on their constituencies, or that some of these pressures are more conductive than others towards success, as success may be defined in that particular society. There is good reason to believe that some of the fundamental characteristics of a class system, as just touched upon, are to be found in all industrial or industrializing societies, including those operating under socialistic régimes that deny the existence of class in their official ideology. But if location in one social stratum as against another has these far-reaching consequences in a society as relatively 'open' as ours, it can easily be seen what the consequences are in more 'closed' systems. We refer here once more to Daniel Lerner's instructive analysis of the traditional societies of the Middle East, in which social location fixed one's identity and one's expectations (even in the imagination) to a degree that most Westerners today find hard even to grasp. Yet European societies before the industrial revolution were not too drastically different, in most of their strata, from Lerner's traditional model. In such societies a man's total being can be gleaned from one look at his social position, as one can glance at a Hindu's forehead and see on it the mark of his caste.

However, even in our own society, superimposed as it were on the class system, there are other stratification systems that are far more rigid and therefore far more determinative of an individual's entire life than that of class. In American society a notable example of this is the racial system, which most sociologists regard as a variety of caste. In such a system an individual's basic social position (that is, his assignment to his caste group) is fixed at birth. At least in theory there is no possibility at all for him to change that position in the course of his lifetime. A man may become as rich as he wants to, he will still be a Negro. Or a man may sink as low as it is possible to sink in terms of the *mores* of society, he will still be white. An individual is born in his caste, must live all his life within it and within all the limitations of conduct that this entails. And, of course, he must marry and procreate within that caste. In actuality, at least in our racial system, there are some possibilities of 'cheating' – namely, the practice of light-skinned Negroes 'passing' as whites. But these possibilities do little to change the total efficacy of the system.

The depressing facts of the American racial system are too well known to necessitate much elaboration here. It is clear that the social location of an individual as a Negro (more so in the South than in the North, of course, but with fewer differences between the two regions than self-righteous

Northern whites commonly allow) implies a far narrower channelling of existential possibilities than happens by way of class. Indeed, the individual's possibilities of class mobility are most definitely determined by his racial location, since some of the most stringent disabilities of the latter are economic in character. Thus a man's conduct, ideas and psychological identity are shaped by race in a manner far more decisive than they commonly are by class.

The constraining force of this location can be seen in its purest form (if such an adjective can be applied, even in a quasi-chemical sense, to so revolting a phenomenon) in the racial etiquette of traditional Southern society, in which every single instance of interaction between members of the two castes was regulated in a stylized ritual designed carefully to honour the one party and to humiliate the other. A Negro risked physical punishment and a white man extreme opprobrium by the slightest deviation from the ritual. Race determined infinitely more than where one could live and with whom one could associate. It determined one's linguistic inflection, one's gestures, one's jokes, and even penetrated one's dreams of salvation. In such a system the criteria of stratification become metaphysical obsessions – as in the case of the Southern gentlewoman who expressed the conviction that her cook was surely bound for the coloured folks' heaven.

A commonly used concept in sociology is that of the definition of the situation. First coined by the American sociologist W. I. Thomas, it means that a social situation *is* what it is defined to be by its participants. In other words, for the sociologist's purposes reality is a matter of definition. This is why the sociologist must analyze earnestly many facets of human conduct that are in themselves absurd or delusional. In the example of the racial system just given, a biologist or physical anthropologist may take one look at the racial beliefs of white Southerners and declare that these beliefs are totally erroneous. He can then dismiss them as but another mythology produced by human ignorance and ill will, pack up his things and go home. The sociologist's task, however, only begins at this point. It does not help at all for him to dismiss the Southern racial ideology as a scientific imbecility. Many social situations are effectively controlled by the definitions of imbeciles. Indeed, the imbecility that defines the situation is part of the stuff of sociological analysis. Thus the sociologist's operational understanding of 'reality' is a somewhat peculiar one, a point to which we shall return again. For the moment it is merely important to point out that the inexorable controls by which social location determines our lives are not done away with by debunking the ideas that undergird these controls.

Nor is this the whole story. Our lives are not only dominated by the inanities of our contemporaries, but also by those of men who have been dead for generations. What is more, each inanity gains credence and reverence with each decade that passes after its original promulgation. As Alfred Schuetz has pointed out, this means that each social situation in which we find ourselves is not only defined by our contemporaries but predefined by our predecessors. Since one cannot possibly talk back to one's ancestors, their ill-conceived constructions are commonly more difficult to get rid of than those built in our own lifetime. This fact is caught in Fontenelle's aphorism that the dead are more powerful than the living.

This is important to stress because it shows us that even in the areas where society apparently allows us some choice the powerful hand of the past narrows down this choice even further. Let us, for example, return to an incident evoked earlier, a scene in which a pair of lovers are sitting in the moonlight. Let us further imagine that this moonlight session turns out to be the decisive one, in which a proposal of marriage is made and accepted. Now, we know that contemporary society imposes considerable limitations on such a choice, greatly facilitating it among couples that fit into the same socio-economic categories and putting heavy obstacles in the way of such as do not. But it is equally clear that even where 'they' who are still alive have made no conscious attempts to limit the choice of the participants in this particular drama, 'they' who are dead have long ago written the script for almost every move that is made. The notion that sexual attraction can be translated into romantic emotion was cooked up by misty-voiced minstrels titillating the imagination of aristocratic ladies about the twelfth century or thereabouts. The idea that a man should fixate his sexual drive permanently and exclusively on one single woman, with whom he is to share bed, bathroom and the boredom of a thousand bleary-eyed breakfasts, was produced by misanthropic theologians some time before that. And the assumption that the initiative in the establishment of this wondrous arrangement should be in the hands of the male, with the female graciously succumbing to the impetuous onslaught of his wooing, goes back right to prehistoric times when savage warriors first descended on some peaceful matriarchal hamlet and dragged away its screaming daughters to their marital cots.

Just as all these hoary ancients have decided the basic framework within which the passions of our exemplary couple will develop, so each step in their courtship has been predefined, prefabricated – if you like, 'fixed'. It is not only that they are supposed to fall in love and to enter into a monogamous marriage in which she gives up her name and he his solvency, that this love must be manufactured at all cost or the marriage will seem insincere to all concerned, and that state and church will watch over the *ménage* with anxious attention once it is established – all of which are fundamental assumptions concocted centuries before ι protagonists were born. Each step in their courtship is laid down in social ritual also and, although there is always some leeway for improvisations, too much adlibbing is likely to risk the success of the whole operation. In this way, our couple progresses predictably (with what a lawyer would call 'due deliberate speed') from movie dates to church dates to meeting-the-family dates, from holding hands to tentative explorations to what they originally planned to save for afterward, from planning their evening to planning their suburban ranch house – with the scene in the moonlight put in its proper place in this ceremonial sequence. Neither of them has invented this game or any part of it. They have only decided that it is with each other, rather than with other possible partners, that they will play it. Nor do they have an awful lot of choice as to what is to happen after the necessary ritual exchange of question and answer. Family, friends, clergy, salesmen of jewelry and of life insurance, florists and interior decorators ensure that the remainder of the game will also be played by the established rules. Nor, indeed, do all these guardians of tradition have to exert much pressure on the principal players, since the expectations of their social world have long ago been built into their own projections of the future – they want precisely that which society expects of them.

If this is so in the most intimate concerns of our existence, it is easy to see that it is the same in almost any social situation encountered in the course of a lifetime. Most of the time the game has been 'fixed' long before we arrive on the scene. All that is left for us to do, most of the time, is to play it with more or with less enthusiasm. The professor stepping in front of his class, the judge pronouncing sentence, the preacher badgering his congregation, the commander ordering his troops into battle – all these are engaged in actions that have been predefined within very narrow limits.

And impressive systems of controls and sanctions stand guard over these limits.

With these considerations behind us, we can now arrive at a more sophisticated understanding of the functioning of social structures. A useful sociological concept on which to base this understanding is that of 'institution'. An institution is commonly defined as a distinctive complex of social actions. Thus we can speak of the law, of class, marriage or organized religion as constituting institutions. Such a definition does not yet tell us in what way the institution relates to the actions of the individuals involved. A suggestive answer to this question has been given by Arnold Gehlen, a contemporary German social scientist. Gehlen conceives of an institution as a regulatory agency, channelling human actions in much the same way as instincts channel animal behaviour. In other words, institutions provide procedures through which human conduct is patterned, compelled to go, in grooves deemed desirable by society. And this trick is performed by making these grooves appear to the individual as the only possible ones.

Let us take an example. Since cats do not have to be taught to chase mice, there is apparently something in a cat's congenital equipment (an instinct, if you like that term) that makes it do so. Presumably, when a cat sees a mouse, there is something in the cat that keeps insisting, Eat! Eat! Eat! The cat does not exactly choose to obey this inner voice. It simply follows the law of its innermost being and takes off after the hapless mouse (which, we suppose, has an inner voice that keeps repeating, Run! Run! Run!). Like Luther, the cat can do no other. But now let us switch back to the couple whose courtship we have previously discussed with such an apparent lack of sympathy. When our young man first beheld the girl destined to provoke the moonlight act (or, if not at first, some time afterward), he also found himself listening to an inner voice uttering a clear imperative. And his subsequent behaviour shows that he also found that imperative irresistible. No, this imperative is *not* what the reader probably just thought of – *that* imperative our young man shares congenitally with young tomcats, chimpanzees and crocodiles, and it does not interest us at the moment. The imperative that concerns us is the one that says to him, Marry! Marry! Marry! For this imperative, unlike the other, our young man was not born with. It was instilled in him by society, reinforced by the countless pressures of family lore, moral education, religion, the mass media and advertising. In other words, marriage is not an instinct but an institution. Yet the way it leads behaviour into predetermined channels is very similar to what the instincts do where they hold sway.

This becomes obvious if we try to imagine what our young man would do in the absence of the institutional imperative. He could, of course, do an almost infinite number of things. He could have sexual relations with the girl, leave and never see her again. Or he could wait until her first child is born and then ask her maternal uncle to raise it. Or he could get together with three buddies of his and ask them whether they would jointly acquire the girl as their common wife. Or he could incorporate her in his harem along with the twenty-three females already living in it. In other words, given his sex drive and his interest in that particular girl, he would be in quite a quandary. Even assuming that he has studied anthropology and knows that all the above-mentioned options are the normal thing to do in some human culture, he would still have a difficult time deciding which one would be the most desirable one to follow in this case. Now we can see what the institutional imperative does for him. It protects him from this quandary. It shuts out all other options in favour of the one that his society has predefined for him. It even bars these other options from his consciousness. It presents him with a formula – to desire is to love is to marry. All that he must do now is to retrace the steps prepared for him in this programme. This may have enough difficulties of its own, but the difficulties are of a very different order from those faced by some proto-male coming up against a proto-female in a clearing of the primeval jungle and having to work out a *modus vivendi* with her. In other words, the situation of marriage serves to channel our young man's conduct, to make him behave according to type. The institutional structure of society supplies the typology for our actions. Only very, very rarely are we in a position to think up new types to pattern ourselves after. Mostly we have a maximum choice between type A and type B, both types predefined for us *a priori*. Thus we might decide to be an artist rather than a businessman. But in either case we shall come up against quite precise predefinitions of the things we must now do. And neither way of life will have been invented by ourselves.

One further aspect of Gehlen's concept of the institution to stress, because it will be significant later on in our argument, is the seeming inevitability of its imperatives. The average young man in our society not only rejects the options of polyandry or polygyny, but, at least for himself, finds them literally unthinkable. He believes that the institutionally predefined course of action is the only one he could possibly take, the only one he is ontologically capable of. Presumably the cat, if it reflected on its persecution of the mouse, would arrive at the same conclusion. The difference is that the cat would be right in its conclusion, while the young man is wrong. As far as we know, a cat that refused to chase mice would be a biological monstrosity, perhaps the result of a malignant mutation, certainly a traitor to the very essence of felinity. But we know very well that having many wives or being one of many husbands is not a betrayal of humanity, in any biological sense, or even of virility. And since it is biologically possible for Arabs to have the one and for Tibetans to be the other, it must also be biologically possible for our young man. Indeed, we know that if the latter had been snatched out of his cradle and shipped to the right alien shores at an early enough age, he would not have grown up to be the red-blooded and more than slightly sentimental all-American boy of our moonlight scene, but would have developed into a lusty polygamist in Arabia or a contented multiple-husband in Tibet. That is, he is deceiving himself (or, more accurately, being deceived by society) when he looks upon his course of action in this matter as inevitable. This means that every institutional structure must depend on deception and all existence in society carries with it an element of bad faith. This glimmering insight may appear thoroughly depressing at first, but, as we shall see, it actually offers us the first glimpse of a view of society somewhat less deterministic than we have obtained so far.

For the moment, however, our considerations of sociological perspective have led us to a point where society looks more like a gigantic Alcatraz than anything else. We have passed from the childish satisfaction at having an address to the grown-up realization that most of the mail is unpleasant. And sociological understanding has only helped us to identify more closely all the personages, dead or alive, who have the privilege of sitting on top of us.

The approach to sociology that comes closest to expressing this sort of view of society is the approach associated with Émile Durkheim and his school. Durkheim emphasized that society is a phenomenon *sui generis*, that is, it confronts us with a massive reality that cannot be reduced to or translated into other terms. He then stated that social facts are 'things', having an objective existence outside of ourselves just like the phenomena of nature. He did this mainly to protect sociology from being swallowed by the imperialistically minded psychologists, but his conception is significant beyond this methodological concern. A 'thing' is something like a rock, for example, that one comes up against, that one cannot move by wishing it out of existence or imagining it as having a different shape. A 'thing' is that against which one can throw oneself in vain, that which is there against all one's desires and hopes, that which can finally fall on one's head and kill one. This is the sense in which society is a collection of 'things'. The law, perhaps more clearly than any other social institution, illustrates this quality of society.

If we follow the Durkheimian conception, then, society confronts us as an objective facticity. It is *there*, something that cannot be denied and that must be reckoned with. Society is external to ourselves. It surrounds us, encompasses our life on all sides. We are *in* society, located in specific sectors of the social system. This location predetermines and predefines almost everything we do, from language to etiquette, from the religious beliefs we hold to the probability that we will commit suicide. Our wishes are not taken into consideration in this matter of social location, and our intellectual resistance to what society prescribes or proscribes avails very little at best, and frequently nothing. Society, as objective and external fact, confronts us especially in the form of coercion. Its institutions pattern our actions and even shape our expectations. They reward us to the extent that we stay within our assigned performances. If we step out of these assignments, society has at its disposal an almost infinite variety of controlling and coercing agencies. The sanctions of society are able, at each moment of existence, to isolate us among our fellow men, to subject us to ridicule, to deprive us of our sustenance and our liberty, and in the last resort to deprive us of life itself. The law and the morality of society can produce elaborate justifications for each one of these sanctions, and most of our fellow men will approve if they are used against us in punishment for our deviance. Finally, we are located in society not only in space but in time. Our society is an historical entity that extends temporally beyond any individual biography. Society antedates us and it will survive us. It was there before we were born and it will be there after we are dead. Our lives are but episodes in its majestic march through time. In sum, society is the walls of our imprisonment in history.

5

SOCIETY IN MAN

IN the preceding chapter we may have given the reader excellent grounds to decide that sociology is ready to take over the title of 'the dismal science' from economics. Having presented the reader with an image of society as a forbidding prison, we ought now to indicate at least some escape tunnels from this gloomy determinism. Before we try to do this, however, we must deepen the gloom a little more.

So far, approaching society mainly under the aspect of its control systems, we have viewed the individual and society as two entities confronting each other. Society has been conceived as an external reality that exerts pressure and coercion upon the individual. If this picture were left unmodified, one would obtain a quite erroneous impression of the actual relationship, namely, an impression of masses of men constantly straining at their leashes, surrendering to the controlling authorities with gnashing teeth, constantly driven to obedience by fear of what may happen to them otherwise. Both common-sense knowledge of society and sociological analysis proper tell us that this is not so. For most of us the yoke of society seems easy to bear. Why? Certainly not because the power of society is less than we indicated in the last chapter. Why then do we not suffer more from this power? The sociological answer to this question has already been alluded to – because most of the time we ourselves desire just that which society expects of us. We *want* to obey the rules. We *want* the parts that society has assigned to us. And this in turn is possible not because the power of society is less, but because it is much more than we have so far asserted. Society not only determines what we do but also what we are. In other words, social location involves our being as well as our conduct. To explicate this crucial element of sociological perspective we shall now look at three further areas of investigation and interpretation, those of role theory, the sociology of knowledge and reference-group theory.

Role theory has been almost entirely an American intellectual development. Some of its germinal insights go back to William James, while its direct parents are two other American thinkers, Charles Cooley and George Herbert Mead. It cannot be our purpose here to give an historical introduction to this quite fascinating portion of intellectual history. Rather than try this even in outline, we shall start more systematically by beginning our consideration of the import of role theory with another look at Thomas's concept of the definition of the situation.

The reader will recall Thomas's understanding of the social situation as a sort of reality agreed upon *ad hoc* by those who participate in it, or, more exactly, those who do the defining of the situation. From the viewpoint of the individual participant this means that each situation he enters confronts him with specific expectations and demands of him specific responses to these expectations. As we have already seen, powerful pressures exist in just about any social situation to ensure that the proper responses are indeed forthcoming. Society can exist by virtue of the fact that most of the time most people's definitions of the most important situations at least coincide approximately. The motives of the publisher and the writer of these lines may be rather different, but the ways the two define the situation in which this book is being produced are sufficiently similar for the joint venture to be possible. In similar fashion there may be quite divergent interests present in a classroom of students, some of them having little connexion with the educational activity that is supposedly going on, but in most cases these interests (say, that one student came to study the subject being taught, while another simply registers for every course taken by a certain redhead he is pursuing) can coexist in the situation without destroying it. In other words, there is a certain amount of leeway in the extent to which response must meet expectation for a situation to remain sociologically viable. Of course, if the definitions of the situation are too widely discrepant, some form of social conflict or disorganization will inevitably result – say, if some students interpret the classroom meeting as a party, or if an author has no intention of producing a book but is using his contract with one publisher to put pressure on another.

While an average individual meets up with very different expectations in different areas of his life in society, the situations that produce these expectations fall into certain clusters. A student may take two courses from two different professors in two different departments, with considerable variations in the expectations met with in the two situations (say, as between formality or informality in the relations between professor and students). Nevertheless, the situations will be sufficiently similar to each other and to other classroom situations previously experienced to enable the student to carry into both situations essentially the same overall response. In other words, in both cases, with but a few modifications, he will be able to *play the role* of student. A role, then, may be defined as a typified response to a typified expectation. Society has predefined the fundamental typology. To use the language of the theatre, from which the concept of role is derived, we can say that society provides the script for all the *dramatis personae*. The individual actors, therefore, need but slip into the roles already assigned to them before the curtain goes up. As long as they play their roles as provided for in this script, the social play can proceed as planned.

The role provides the pattern according to which the

individual is to act in the particular situation. Roles, in society as in the theatre, will vary in the exactness with which they lay down instructions for the actor. Taking occupational roles for an instance, a fairly minimal pattern goes into the role of garbage collector, while physicians or clergymen or officers have to acquire all kinds of distinctive mannerisms, speech and motor habits, such as military bearing, sanctimonious diction or bedside cheer. It would, however, be missing an essential aspect of the role if one regarded it merely as a regulatory pattern for externally visible actions. One feels more ardent by kissing, more humble by kneeling and more angry by shaking one's fist. That is, the kiss not only expresses ardour but manufactures it. Roles carry with them both certain actions and the emotions and attitudes that belong to these actions. The professor putting on an act that pretends to wisdom comes to feel wise. The preacher finds himself believing what he preaches. The soldier discovers martial stirrings in his breast as he puts on his uniform. In each case, while the emotion or attitude may have been present before the role was taken on, the latter inevitably strengthens what was there before. In many instances there is every reason to suppose that nothing at all anteceded the playing of the role in the actor's consciousness. In other words, one becomes wise by being appointed a professor, believing by engaging in activities that presuppose belief, and ready for battle by marching in formation.

Let us take an example. A man recently commissioned as an officer, especially if he came up through the ranks, will at first be at least slightly embarrassed by the salutes he now receives from the enlisted men he meets on his way. Probably he will respond to them in a friendly, almost apologetic manner. The new insignia on his uniform are at that point still something that he has merely put on, almost like a disguise. Indeed, the new officer may even tell himself and others that underneath he is still the same person, that he simply has new responsibilities (among which, *en passant*, is the duty to accept the salutes of enlisted men). This attitude is not likely to last very long. In order to carry out his new role of officer, our man must maintain a certain bearing. This bearing has quite definite implications. Despite all the double-talk in this area that is customary in so-called democratic armies, such as the American one, one of the fundamental implications is that an officer is a superior somebody, entitled to obedience and respect on the basis of this superiority. Every military salute given by an inferior in rank is an act of obeisance, received as a matter of course by the one who returns it. Thus, with every salute given and accepted (along, of course, with a hundred other ceremonial acts that enhance his new status) our man is fortified in his new bearing – and in its, as it were, ontological presuppositions. He not only acts like an officer, he feels like one. Gone are the embarrassment, the apologetic attitude, the I'm-just-another-guy-really grin. If on some occasion an enlisted man should fail to salute with the appropriate amount of enthusiasm or even commit the unthinkable act of failing to salute at all, our officer is not merely going to punish a violation of military regulations. He will be driven with every fibre of his being to redress an offence against the appointed order of his cosmos.

It is important to stress in this illustration that only very rarely is such a process deliberate or based on reflection. Our man has not sat down and figured out all the things that ought to go into his new role, including the things that he ought to feel and believe. The strength of the process comes precisely from its unconscious, unreflecting character. He has become an officer almost as effortlessly as he grew into a person with blue eyes, brown hair and a height of six feet. Nor would it be correct to say that our man must be rather stupid and quite an exception among his comrades. On the contrary, the exception is the man who reflects on his roles and his role changes (a type, by the way, who would probably make a poor officer). Even very intelligent people, when faced with doubt about their roles in society, will involve themselves even more in the doubted activity rather than withdraw into reflection. The theologian who doubts his faith will pray more and increase his church attendance, the businessman beset by qualms about his rat-race activities starts going to the office on Sundays too, and the terrorist who suffers from nightmares volunteers for nocturnal executions. And, of course, they are perfectly correct in this course of action. Each role has its inner discipline, what Catholic monastics would call its 'forma-tion'. The role forms, shapes, patterns both action and actor. It is very difficult to pretend in this world. Normally, one becomes what one plays at.

Every role in society has attached to it a certain identity. As we have seen, some of these identities are trivial and temporary ones, as in some occupations that demand little modification in the being of their practitioners. It is not difficult to change from garbage collector to night watch-man. It is considerably more difficult to change from clergyman to officer. It is very, very difficult to change from Negro to white. And it is almost impossible to change from man to woman. These differences in the ease of role chang-ing ought not to blind us to the fact that even identities that we consider to be our essential selves have been socially assigned. Just as there are racial roles to be acquired and identified with, so there are sexual roles. To say 'I am a man' is just as much a proclamation of role as to say 'I am a colonel in the U.S. Army'. We are well aware of the fact that one is born a male, while not even the most humourless martinet imagines himself to have been born with a golden eagle sitting on his umbilical cord. But to be biologically male is a far cry from the specific, socially defined (and, of course, socially relative) role that goes with the statement 'I am a man'. A male child does not have to learn to have an erection. But he must learn to be aggressive, to have ambitions, to compete with others, and to be suspicious of too much gentleness in himself. The male role in our society, however, requires all these things that one must learn, as does a male identity. To have an erection is not enough – if it were, regiments of psychotherapists would be out of work.

This significance of role theory could be summarized by saying that, in a sociological perspective, identity is socially bestowed, socially sustained and socially transformed. The example of the man in process of becoming an officer may suffice to illustrate the way in which identities are bestowed in adult life. However, even roles that are much more fundamentally part of what psychologists would call our personality than those associated with a particular adult activity are bestowed in very similar manner through a social process. This has been demonstrated over and over again in studies of so-called socialization – the process by which a child learns to be a participant member of society.

Probably the most penetrating theoretical account of this process is the one given by Mead, in which the genesis of the self is interpreted as being one and the same event as the discovery of society. The child finds out who he is as he learns what society is. He learns to play roles properly belonging to him by learning, as Mead put it, 'to take the role of the other' – which, incidentally, is the crucial sociopsychological function of play, in which children masquerade with a variety of social roles and in doing so discover the significance of those being assigned to them. All this learning occurs, and can only occur, in interaction with other human beings, be it the parents or whoever else raises the child. The child first takes on roles *vis-à-vis* what Mead calls his 'significant others', that is, those persons who deal with him intimately and whose attitudes are decisive for the formation of his conception of himself. Later, the child learns that the roles he plays are not only relevant to this intimate circle, but relate to the expectations directed toward him by society at large. This higher level of abstraction in the social response Mead calls the discovery of the 'generalized other'. That is, not only the child's mother expects him to be good, clean and truthful, society in general does so as well. Only when this general conception of society emerges is the child capable of forming a clear conception of himself. 'Self' and 'society', in the child's experience, are the two sides of the same coin.

In other words, identity is not something 'given', but is bestowed in acts of social recognition. We become that as which we are addressed. The same idea is expressed in Cooley's well-known description of the self as a reflection in a looking glass. This does not mean, of course, that there are not certain characteristics an individual is born with, that are carried by his genetic heritage regardless of the social environment in which the latter will have to unfold itself. Our knowledge of man's biology does not as yet allow us a very clear picture of the extent to which this may be true. We do know, however, that the room for social formation within those genetic limits is very large indeed. Even with the biological questions left largely unsettled, we can say that to be human is to be recognized as human, just as to be a certain kind of man is to be recognized as such. The child deprived of human affection and attention becomes dehumanized. The child who is given respect comes to respect himself. A little boy considered to be a *schlemiel* becomes one, just as a grown-up treated as an awe-inspiring young god of war begins to think of himself and act as is appropriate to such a figure – and, indeed, merges his identity with the one he is presented with in these expectations.

Identities are socially bestowed. They must also be socially sustained, and fairly steadily so. One cannot be human all by oneself and, apparently, one cannot hold on to any particular identity all by oneself. The self-image of the officer as an officer can be maintained only in a social context in which others are willing to recognize him in this identity. If this recognition is suddenly withdrawn, it usually does not take very long before the self-image collapses.

Cases of radical withdrawal of recognition by society can tell us much about the social character of identity. For example, a man turned overnight from a free citizen into a convict finds himself subjected at once to a massive assault on his previous conception of himself. He may try desperately to hold on to the latter, but in the absence of others in his immediate environment confirming his old identity he will find it almost impossible to maintain it within his own consciousness. With frightening speed he will discover that he is acting as a convict is supposed to, and feeling all the things that a convict is expected to feel. It would be a misleading perspective on this process to look upon it simply as one of the disintegration of personality. A more accurate way of seeing the phenomenon is as a reintegration of personality, no different in its socio-psychological dynamics from the process in which the old identity was integrated. It used to be that our man was treated by all the important people around him as responsible, dignified, considerate and aesthetically fastidious. Consequently he was able to be all these things. Now the walls of the prison separate him from those whose recognition sustained him in the exhibition of these traits. Instead he is now surrounded by people who treat him as irresponsible, swinish in behaviour, only out for his own interests and careless of his appearance unless forced to take care by constant supervision. The new expectations are typified in the convict role that responds to them just as the old ones were integrated into a different pattern of conduct. In both cases, identity comes with conduct and conduct occurs in response to a specific social situation.

Extreme cases in which an individual is radically stripped of his old identity simply illustrate more sharply processes that occur in ordinary life. We live our everyday lives within a complex web of recognitions and non-recognitions. We work better when we are given encouragement by our superiors. We find it hard to be anything but clumsy in a gathering where we know people have an image of us as awkward. We become wits when people expect us to be funny, and interesting characters when we know that such a reputation has preceded us. Intelligence, humour, manual skills, religious devotion and even sexual potency respond with equal alacrity to the expectations of others. This makes understandable the previously mentioned process by which individuals choose their associates in such a way that the latter sustain their self-interpretations. To put this succinctly, every act of social affiliation entails a choice of identity. Conversely every identity requires specific social affiliations for its survival. Birds of the same feather flock together not as a luxury but out of necessity. The intellectual becomes a slob after he is kidnapped by the army. The theological student progressively loses his sense of humour as he approaches ordination. The worker who breaks all norms finds that he breaks even more after he has been given a medal by management. The young man with anxieties about his virility becomes hell-on-wheels in bed when he finds a girl who sees him as an avatar of Don Giovanni.

To relate these observations to what was said in the last chapter, the individual locates himself in society within systems of social control, and every one of these contains an identity-generating apparatus. Insofar as he is able the individual will try to manipulate his affiliations (and especially his intimate ones) in such a way as to fortify the identities that have given him satisfaction in the past — marrying a girl who thinks he has something to say, choosing friends who regard him as entertaining, selecting an occupation that gives him recognition as up-and-coming. In many cases, of course, such manipulation is not possible. One must then do the best one can with the identities one is thrown.

Such sociological perspective on the character of identity gives us a deeper understanding of the human meaning of prejudice. As a result, we obtain the chilling perception that the prejudging not only concerns the victim's external fate at the hands of his oppressors, but also his consciousness as it is shaped by their expectations. The most terrible thing that prejudice can do to a human being is to make him tend to become what the prejudiced image of him says that he is. The Jew in an anti-Semitic milieu must struggle hard not to become more and more like the anti-Semitic stereotype, as must the Negro in a racist situation. Significantly, this struggle will only have a chance of success when the individual is protected from succumbing to the prejudiced programme for his personality by what we could call the counter-recognition of those within his immediate community. The Gentile world might recognize him as but another despicable Jew of no consequence, and treat him accordingly, but this non-recognition of his worth may be balanced by the counter-recognition of him within the Jewish community itself as, say, the greatest Talmudic scholar in Latvia.

In view of the socio-psychological dynamics of this deadly game of recognitions, it should not surprise us that the problem of 'Jewish identity' arose only among modern Western Jews when assimilation into the surrounding Gentile society had begun to weaken the power of the Jewish community itself to bestow alternate identities on its members as against the identities assigned to them by anti-Semitism. As an individual is forced to gaze at himself in a mirror so constructed as to let him see a leering monster, he must frantically search for other men with other mirrors, unless he is to forget that he ever had another face. To put this a little differently, human dignity is a matter of social permission.

The same relationship between society and identity can be seen in cases where, for one reason or another, an individual's identity is drastically changed. The transformation of identity, just as its genesis and its maintenance, is a social process. We have already indicated the way in which any reinterpretation of the past, any 'alternation' from one self-image to another, requires the presence of a group that conspires to bring about the metamorphosis. What anthropologists call a rite of passage involves the repudiation of an old identity (say, that of being a child) and the initiation into a new one (such as that of adult). Modern societies have milder rites of passage, as in the institution of the engagement, by which the individual is gently led by a general conspiracy of all concerned over the threshold between bachelor freedom and the captivity of marriage. If it were not for this institution, many more would panic at the last moment before the enormity of what they are about to undertake.

We have also seen how 'alternation' operates to change identities in such highly structured situations as religious training or psychoanalysis. Again taking the latter as a timely illustration, it involves an intensive social situation in which the individual is led to repudiate his past conception of himself and to take on a new identity, the one that has been programmed for him in the psychoanalytic ideology. What psychoanalysts call 'transference', the intense social relationship between analyst and analysand, is essentially the creation of an artificial social milieu within which the alchemy of transformation can occur, that is, within which this alchemy can become plausible to the individual. The longer the relationship lasts and the more intensive it becomes, the more committed does the individual become to his new identity. Finally, when he is 'cured', this new identity has indeed become what he is. It will not do, therefore, to dismiss with a Marxist guffaw the psychoanalyst's claim that his treatment is more effective if the patient sees him frequently, does so over a long time and pays a considerable fee. While it is obviously in the analyst's economic interest to hold to this position, it is quite plausible sociologically that the position is factually correct. What is actually 'done' in psychoanalysis is that a new identity is constructed. The individual's commitment to this new identity will obviously increase the more intensively, the longer and the more painfully he invests in its manufacture. Certainly his capacity to reject the whole business as a fake has become rather minimal after an investment of several years of his life and thousands of dollars of hard-earned cash.

The same kind of 'alchemistic' environment is established in situations of 'group therapy'. The recent popularity of the latter in American psychiatry can again not be interpreted simply as an economic rationalization. It has its sociological basis in the perfectly correct understanding that group pressures work effectively to make the individual accept the new mirror-image that is being presented to him. Erving Goffman, a contemporary sociologist, has given us a vivid description of how these pressures work in the context of a mental hospital, with the patients finally 'selling out' to the psychiatric interpretation of their existence that is the common frame of reference of the 'therapeutic' group.

The same process occurs whenever an entire group of individuals is to be 'broken' and made to accept a new definition of themselves. It happens in basic training for draftees in the army; much more intensively in the training of personnel for a permanent career in the army, as at military academies. It happens in the indoctrination and 'formation' programmes of cadres for totalitarian organizations, such as the Nazi SS or the Communist Party *élite*. It has happened for many centuries in monastic novitiates. It has recently been applied to the point of scientific precision in the 'brainwashing' techniques employed against prisoners of totalitarian secret police organizations. The violence of such procedures, as compared with the more routine initiations of society, is to be explained sociologically in terms of the radical degree of transformation of identity that is sought and the functional necessity in these cases that commitment to the transformed identity be foolproof against new 'alternations'.

Role theory, when pursued to its logical conclusions, does far more than provide us with a convenient shorthand for the description of various social activities. It gives us a sociological anthropology, that is, a view of man based on his existence in society. This view tells us that man plays dramatic parts in the grand play of society, and that, speaking sociologically, he *is* the masks that he must wear to do so. The human person also appears now in a dramatic context, true to its theatrical etymology (*persona,* the technical term given to the actors' masks in classical theatre). The person is perceived as a repertoire of roles, each one properly equipped with a certain identity. The range of an individual person can be measured by the number of roles he is capable of playing. The person's biography now appears to us as an uninterrupted sequence of stage performances, played to different audiences, sometimes involving drastic changes of costume, always demanding that the actor *be* what he is playing.

Such a sociological view of personality is far more radical in its challenge to the way that we commonly think of ourselves than most psychological theories. It challenges radically one of the fondest presuppositions about the self – its continuity. Looked at sociologically, the self is no longer a solid, given entity that moves from one situation to another. It is rather a process, continuously created and re-created in each social situation that one enters, held together by the slender thread of memory. How slender this thread is, we have seen in our discussion of the reinterpretation of the past. Nor is it possible within this framework of understanding to take refuge in the unconscious as containing the 'real' contents of the self, because the presumed unconscious self is just as subject to social production as is the so-called conscious one, as we have seen. In other words, man is not *also* a social being, but he is social in every aspect of his being that is open to empirical investigation. Still speaking sociologically, then, if one wants to ask who an individual 'really' is in this kaleidoscope of roles and identities, one can answer only by enumerating the situations in which he is one thing and those in which he is another.

Now, it is clear that such transformations cannot occur *ad infinitum* and that some are easier than others. An individual becomes so habituated to certain identities that, even when his social situation changes, he has difficulty keeping up with the expectations newly directed toward him. The difficulties that healthy and previously highly active individuals have when they are forced to retire from their occupation show this very clearly. The transformability of the self depends not *only* on its social context, but also on the degree of its habituation to previous identities and perhaps also on certain genetically given traits. While these modifications in our model are necessary to avoid a radicalization of our position, they do not detract appreciably from the discontinuity of the self as revealed by sociological analysis.

If this not very edifying anthropological model is reminiscent of any other, it would be of that employed in early Buddhist psychology in India, in which the self was compared to a long row of candles, each of which lights the wick of its neighbour and is extinguished in that moment. The Buddhist psychologists used this picture to decry the Hindu notion of the transmigration of the soul, meaning to say thereby that there is no entity that passes from one candle to another. But the same picture fits our present anthropological model quite well.

One might obtain the impression from all of this that there is really no essential difference between most people and those afflicted with what psychiatry calls 'multiple personality'. If someone wanted to harp on the word 'essential' here, the sociologist might agree with the statement. The actual difference, however, is that for 'normal' people (that is, those so recognized by their society) there are strong pressures toward consistency in the various roles they play and the identities that go with these roles. These pressures are both external and internal. Externally the others with whom one must play one's social games, and on whose recognition one's own parts depend, demand that one present at least a relatively consistent picture to the world. A certain degree of role discrepancy may be permitted, but if certain tolerance limits are passed society will withdraw its recognition of the individual in question, defining him as a moral or psychological aberration. Thus society will allow an individual to be an emperor at work and a serf at home, but it will not permit him to impersonate a police officer or to wear the costume assigned to the other sex. In order to stay within the limits set to his masquerades, the individual may have to resort to complicated manoeuvres to make sure that one role remains segregated from the other. The imperial role in the office is endangered by the appearance of one's wife at a directors' meeting, or one's role in one circle as an accomplished *raconteur* is threatened by the intrusion of someone from that other circle in which one has been typed as the fellow who never opens his mouth without putting his foot into it. Such role segregation is increasingly possible in our contemporary urban civilization, with its anonymity and its means of rapid transportation, although even here there is a danger that people with contradictory images of oneself may suddenly bump into each other and endanger one's whole stage management. Wife and secretary might meet for coffee, and between them reduce both home-self and office-self to a pitiable shambles. At that point, for sure, one will require a psychotherapist to put a new Humpty Dumpty together again.

There are also internal pressures toward consistency, possibly based on very profound psychological needs to perceive oneself as a totality. Even the contemporary urban masquerader, who plays mutually irreconcilable roles in different areas of his life, may feel internal tensions though he can successfully control external ones by carefully segregating his several *mises en scène* from each other. To avoid such anxieties people commonly segregate their consciousness as well as their conduct. By this we do not mean that they 'repress' their discrepant identities into some 'unconscious', for within our model we have every reason to be suspicious of such concepts. We rather mean that they focus their attention only on that particular identity that, so to speak, they require at the moment. Other identities are forgotten for the duration of this particular act. The way in which socially disapproved sexual acts or morally questionable acts of any kind are segregated in consciousness may serve to illustrate this process. The man who engages in, say, homosexual masochism has a carefully constructed identity set aside for just these occasions. When any given occasion is over, he checks that identity again at the gate, so to speak, and returns home as affectionate father, responsible husband, perhaps even ardent lover of his wife. In the

same way, the judge who sentences a man to death seg-
regates the identity in which he does this from the rest of
his consciousness, in which he is a kindly, tolerant and
sensitive human being. The Nazi concentration-camp commander who writes sentimental letters to his children is but
an extreme case of something that occurs in society all the
time.

It would be a complete misunderstanding of what has just
been said if the reader now thought that we are presenting a
picture of society in which everybody schemes, plots and
deliberately puts on disguises to fool his fellow men. On the
contrary, role-playing and identity-building processes are
generally unreflected and unplanned, almost automatic. The
psychological needs for consistency of self-image just mentioned ensure this. Deliberate deception requires a degree of
psychological self-control that few people are capable of.
That is why insincerity is rather a rare phenomenon. Most
people are sincere, because this is the easiest course to take
psychologically. That is, they believe in their own act, con-
veniently forget the act that preceded it, and happily go
through life in the conviction of being responsible in all its
demands. Sincerity is the consciousness of the man who is
taken in by his own act. Or as it has been put by David
Riesman, the sincere man is the one who believes in his own
propaganda. In view of the socio-psychological dynamics
just discussed, it is much more likely that the Nazi
murderers are sincere in their self-portrayals as having been
bureaucrats faced with certain unpleasant exigencies that
actually were distasteful to them than to assume that they
say this only in order to gain sympathy from their judges.
Their humane remorse is probably just as sincere as their
erstwhile cruelty. As the Austrian novelist Robert Musil has
put it, in every murderer's heart there is a spot in which he
is eternally innocent. The seasons of life follow one another,
and one must change one's face as one changes one's clothes.

At the moment we are not concerned with the psychological difficulties or the ethical import of such 'lack of character'. We only want to stress that it is the customary procedure.

To tie up what has just been said about role theory with what was said in the preceding chapter about control systems we refer to what Hans Gerth and C. Wright Mills have called 'person selection'. Every social structure selects those persons that it needs for its functioning and eliminates in one way or another those that do not fit. If no persons are available to be selected, they will have to be invented – or rather, they will be produced in accordance with the required specifications. In this way, through its mechanisms of socialization and 'formation', society manufactures the personnel it requires to keep going. The sociologist stands on its head the common-sense idea that certain institutions arise because there are certain persons around. On the contrary, fierce warriors appear because there are armies to be sent out, pious men because there are churches to be built, scholars because there are universities to be staffed, and murderers because there are killings to be performed. It is not correct to say that each society gets the men it deserves. Rather, each society produces the men it needs. We can derive some comfort from the fact that this production process sometimes runs into technical difficulties. We shall see later that it can also be sabotaged. For the moment, however, we can see that role theory and its concomitant perceptions add an important dimension to our sociological perspective on human existence.

If role theory provides us with vivid insights into the presence of society in man, similar insights from a very different starting point can be obtained from the so-called sociology of knowledge. Unlike role theory, the sociology of knowledge is European in origin. The term was first coined in the 1920s by the German philosopher Max Scheler. Another European scholar, Karl Mannheim, who spent the closing years of his life in England, is largely responsible for bringing the new discipline to the attention of Anglo-Saxon thought. This is not the place to go into the very intriguing intellectual ancestry of the sociology of knowledge, which includes Marx, Nietzsche and German historicism. The way the sociology of knowledge fits into our argument is by showing us that ideas as well as men are socially located. And this, indeed, can serve as a definition of the discipline for our purposes – the sociology of knowledge concerns itself with the social location of ideas.

The sociology of knowledge, more clearly than any other branch of sociology, makes clear what is meant by saying that the sociologist is the guy who keeps asking '*Says who?* '. It rejects the pretence that thought occurs in isolation from the social context within which particular men think about particular things. Even in the case of very abstract ideas that seemingly have little social connexion, the sociology of knowledge attempts to draw the line from the thought to the thinker to his social world. This can be seen most easily in those instances when thought serves to legitimate a particular social situation, that is, when it explains, justifies and sanctifies it.

Let us construct a simple illustration. Let us assume that in a primitive society some needed foodstuff can be obtained only by travelling to where it grows through treacherous, shark-infested waters. Twice every year the men of the tribe set out in their precarious canoes to get this food. Now, let us assume that the religious beliefs of this society contain an article of faith that says that every man who fails to go on this voyage will lose his virility, except for the priests, whose virility is sustained by their daily sacrifices to the gods. This belief provides a motivation for those who expose themselves to the dangerous journey and simultaneously a legitimation for the priests who regularly stay at home. Needless to add, we will suspect in this example that it was the priests who cooked up the theory in the first place. In other words, we will assume that we have here a priestly ideology. But this does not mean that the latter is not functional for the society as a whole – after all, somebody must go or there will be starvation.

We speak of an ideology when a certain idea serves a vested interest in society. Very frequently, though not always, ideologies systematically distort social reality in order to come out where it is functional for them to do so. In looking at the control systems set up by occupational groups we have already seen the way in which ideologies can legitimate the activities of such groups. Ideological thinking, however, is capable of covering much larger human collectivities. For example, the racial mythology of the American South serves to legitimate a social system practised by millions of human beings. The ideology of 'free enterprise' serves to camouflage the monopolistic practices of large American corporations whose only common characteristic with the old-style entrepreneur is a steadfast readiness to defraud the public. The Marxist ideology, in turn, serves to legitimate the tyranny practised by the Communist Party apparatus whose interests have about as much in common with Karl Marx's as those of Elmer Gantry had with the Apostle Paul's. In each case, the ideology both justifies what is done by the group whose vested interest is served and interprets social reality in such a way that the justification is made plausible. This interpretation often appears bizarre to an outsider who 'does not understand the problem' (that is, who does not share the vested interest). The Southern racist must simultaneously maintain that white women have a profound revulsion at the very thought of sexual relations with a Negro and that the slightest inter-racial sociability will straightway lead to such sexual relations. And the corporation executive will maintain that his activities to fix prices are undertaken in defence of a free market. And the Communist Party official will have a way of explaining that the limitation of electoral choice to candidates approved by the party is an expression of true democracy.

It should be stressed again in this connexion that commonly the people putting forth these propositions are perfectly sincere. The moral effort to lie deliberately is beyond most people. It is much easier to deceive oneself. It is, therefore, important to keep the concept of ideology distinct from notions of lying, deception, propaganda or legerdemain. The liar, by definition, knows that he is lying. The ideologist does not. It is not our concern at this point to ask which of the two is ethically superior. We only stress once more the unreflected and unplanned way in which society normally operates. Most theories of conspiracy grossly over-estimate the intellectual foresight of the conspirators.

Ideologies can also function 'latently', to use Merton's expression in another context. Let us return another time to the American South as an example. One interesting fact about it is the geographical coincidence between the Black Belt and the Bible Belt. That is, roughly the same area that practises the Southern racial system in pristine purity also has the heaviest concentration of ultra-conservative, fundamentalist Protestantism. This coincidence can be explained historically, by pointing to the isolation of Southern Protestantism from broader currents of religious thought ever since the great denominational splits over slavery in the antebellum period. The coincidence could also be interpreted as expressing two different aspects of intellectual barbarism. We would not quarrel with either explanation, but would contend that a sociological interpretation in terms of ideological functionality will carry us farther in understanding the phenomenon.

Protestant fundamentalism, while it is obsessed with the

idea of sin, has a curiously limited concept of its extent. Revivalistic preachers thundering against the wickedness of the world invariably fasten on a rather limited range of moral offences – fornication, drink, dancing, gambling, swearing. Indeed, so much emphasis is placed on the first of these that, in the *lingua franca* of Protestant moralism, the term 'sin' itself is almost cognate with the more specific term 'sexual offence'. Whatever one may say otherwise about this catalogue of pernicious acts, they all have in common their essentially *private* character. Indeed, when a revivalistic preacher mentions public matters at all, it is usually in terms of the private corruption of those holding public offices. Government officials steal, which is bad. They also fornicate, drink, and gamble, which is presumably even worse. Now, the limitation to private wrong-doing in one's concept of Christian ethics has obvious functions in a society whose central social arrangements are dubious, to say the least, when confronted with certain teachings of the New Testament and with the egalitarian creed of the nation that considers itself to have roots in the same. Protestant fundamentalism's private concept of morality thus concentrates attention on those areas of conduct that are irrelevant to the maintenance of the social system, and diverts attention from those areas where ethical inspection would create tensions for the smooth operation of the system. In other words, Protestant fundamentalism is ideologically functional in maintaining the social system of the American South. We need not go on to the point where it directly legitimates the system, as when segregation is proclaimed as a God-given natural order. But even in the absence of such 'manifest' legitimation, the religious beliefs in question function 'latently' to keep the system going.

While the analysis of ideologies illustrates sharply what is meant by the social location of ideas, it is still too narrow in scope to demonstrate the full significance of the sociology of knowledge. The latter discipline is not concerned solely with ideas that serve vested interests or that distort social reality. The sociology of knowledge rather regards the entire realm of thought as its province, not of course by regarding itself as the arbiter of validity (which would be megalomaniac), but insofar as thought of any kind is grounded in society. This does not mean (as it would in a Marxist interpretation) that all human thought is regarded as a direct 'reflection' of social structure, nor does it mean that ideas are regarded as completely impotent in shaping the course of events. But it does mean that all ideas are scrutinized carefully to see their location in the social exist-ence of those who thought them. To that extent, at any rate, it is correct to say that the sociology of knowledge is anti-idealistic in its tendency.

Each society can be viewed in terms of its social structure and its socio-psychological mechanisms, and also in terms of the world view that serves as the common universe inhabited by its members. World views vary socially, from one society to another and within different segments of the same society. It is in this sense that one says that a Chinese 'lives in a different world' from that of a Westerner. To stay with this example for an instant, Marcel Granet, a French sinologist heavily influenced by Durkheimian socio-logy, has analyzed Chinese thought in just this perspective of bringing out its 'different world'. The difference, of course, is patent in such matters as political philosophy, religion or ethics. But Granet argued that fundamental differences could also be found in such categories as time, space and number. Very similar assertions have been made in other analyses of this kind, as those comparing the 'worlds' of ancient Greece and ancient Israel, or the 'world' of traditional Hinduism with that of the modern West.

The sociology of religion is one of the most fruitful areas for this kind of investigation, partly perhaps because the paradox of social location appears here in particularly cogent form. It seems quite inappropriate that ideas concerning the gods, the cosmos and eternity should be located in the social systems of men, bound to all the human relativities of geography and history. This has been one of the emotional stumbling blocks in Biblical scholarship, especially as the latter has tried to find what it calls the *Sitz im Leben* (literally the 'site in life' – pretty much what we have called social location) of particular religious phenomena. It is one thing to debate the timeless assertions of Christian faith, quite another to investigate how those assertions may relate to the very timely frustrations, ambitions or resentments of particular social strata in the polyglot cities of the Roman Empire into which the first Christian missionaries carried their message. But more than that, the very phenomenon of religion as such can be socially located in terms of specific functions, such as its legitimation of political authority or its assuagement of social rebellion (what Weber called the 'theodicy of suffering' – that is, the way in which religion gives meaning to suffering, thereby changing it from a source of revolution to a vehicle of redemption). The universality of religion, far from being a proof of its metaphysical validity, is explicable in terms of such social functions. Moreover, the changes in religious patterns in the course of history can also be interpreted in sociological terms.

Take as an example the distribution of religious allegiances in the contemporary Western world. In many Western countries church attendance can be correlated almost perfectly with class affiliation, so that, for instance, religious activity is one of the marks of middle-class status, while abstinence from such activity characterizes the working class. In other words, there appears to be a relationship between one's faith in, say, the Trinity (or at least outward expression of this faith) and one's annual income – below a certain income level it would seem that such faith loses all plausibility, while above that level it becomes a matter of course. The sociology of knowledge will ask how this sort of relationship between statistics and salvation has come about. The answers, inevitably, will be sociological ones – in terms of the functionality of religion in this or that social milieu. Naturally the sociologist will not be able to make any statements about theological questions in themselves, but he will be able to show that these questions have rarely been negotiated in a social vacuum.

To return to a previous illustration, the sociologist will not be able to advise people as to whether they should commit themselves to Protestant fundamentalism or to a less conservative version of that faith, but he will be able to demonstrate to them how either choice will function socially. Nor will he be able to decide for people whether they should have their infants baptized or wait until a later time for this act, but he will be in a position to inform them which expectation they will encounter in which social stratum. Nor can he even make a suggestion as to whether people should anticipate a life after death, but he can tell them which careers in this life will make it advisable for them to pretend at least to such belief.

Beyond these questions of the social distribution of religiosity, some contemporary sociologists (for example, Helmut Schelsky and Thomas Luckmann) have raised the question whether the personality types produced by modern industrial civilization permit the continuation of traditional religious patterns at all and whether, for various sociological and socio-psychological reasons, the Western world may not already be in a post-Christian stage. To pursue these questions, however, would take us beyond our argument. The religious illustrations may have been sufficient to indicate in what way the sociology of knowledge locates ideas in society.

The individual, then, derives his world view socially in very much the same way that he derives his roles and his identity. In other words, his emotions and his self-interpretation like his actions are predefined for him by society, and so is his cognitive approach to the universe that surrounds him. This fact Alfred Schuetz has caught in his phrase 'world-taken-for-granted' – the system of apparently self-evident and self-validating assumptions about the world that each society engenders in the course of its history. This socially determined world view is, at least in part, already given in the language used by the society. Certain linguists may have exaggerated the importance of this factor alone in creating any given world view, but there can be little doubt that one's language at least helps to shape one's relationship to reality. And, of course, our language is not chosen by ourselves but imposed upon us by the particular social group that is in charge of our initial socialization. Society predefines for us that fundamental symbolic apparatus with which we grasp the world, order our experience and interpret our own existence.

In the same way, society supplies our values, our logic and the store of information (or, for that matter, misinformation) that constitutes our 'knowledge'. Very few people, and even they only in regard to fragments of this world view, are in a position to re-evaluate what has thus been imposed on them. They actually feel no need for reappraisal because the world view into which they have been socialized appears self-evident to them. Since it is also so regarded by almost everyone they are likely to deal with in their own society, the world view is self-validating. Its 'proof' lies in the reiterated experience of other men who take it for granted also. To put this perspective of the sociology of knowledge into one succinct proposition: Reality is socially constructed. In this proposition, the sociology of knowledge helps to round out Thomas's statement on the power of social definition and throws further light on the sociological picture of the precarious nature of reality.

Role theory and the sociology of knowledge represent very different strands in sociological thought. Their crucial insights into social processes have not really been integrated as yet theoretically, except perhaps in the important contemporary sociological system of Talcott Parsons which is too complex to discuss within the present argument. A relatively simple nexus between the two approaches, however, is furnished by so-called reference-group theory, once more an American development in the field. First used by Herbert Hyman in the 1940s, the concept of the reference group has been developed further by a number of American sociologists (among whom Robert Merton and Tamotsu Shibutani are especially important). It has been very useful in research into the functioning of organizations of various kinds, such as military or industrial organizations, though this use does not interest us here.

A distinction has been made between reference groups of which one is a member and those towards which one orients one's actions. The latter variety will serve us here. A reference group, in this sense, is the collectivity whose opinions, convictions and courses of action are decisive for the formation of our own opinions, convictions and courses of action. The reference group provides us with a model with which we can continually compare ourselves. Specifically, it gives us a particular slant on social reality, one that may or may not be ideological in the aforementioned sense, but that will in any case be part and parcel of our allegiance to this particular group.

Some time ago there was a cartoon in *The New Yorker* showing a neatly dressed young collegian speaking to an unkempt girl marching in a procession with a sign demanding that bomb tests be banned. The caption went something like this: 'I guess this means that I won't see you at the

Young Conservatives Club tonight'. This little vignette aptly shows the choice of reference groups that may be available to a college student today. Any college of more than minimum size will offer a considerable variety of such groups to choose from. The affiliation-hungry student can join any number of politically defined groups, or he can orient himself towards a beatnik crowd, or he can latch on to a circle of swift socialites, or he can simply hang around with the coterie that has formed itself around a popular professor of English. It goes without saying that, in each case, there will be certain requirements to be met in terms of his dress and demeanour – sprinkling his conversation with left-wing jargon, or boycotting the local barber's shop, or wearing button-down collars and repressed ties, or going barefoot after mid-March. But the choice of group will carry with it an assortment of intellectual symbols as well, symbols that had better be exhibited with a show of commitment – reading *National Review* or *Dissent* (as the case may be), enjoying Allen Ginsberg read to the accompaniment of the farthest-out-possible jazz, knowing the first names of the presidents of a string of corporations one has one's eye on, or evincing unspeakable contempt for anyone admitting ignorance of the Metaphysical Poets. Goldwater Republicanism, Trotskyism, Zen or New Criticism – all these august possibilities of *Weltanschauung* can make or mar one's dates on Saturday evenings, poison one's relationships with room-mates or become the basis of staunch alliances with individuals one previously avoided like the bubonic plague. And one then discovers that one can make some girls with a sportscar and some with John Donne. Naturally, only an evil-minded sociologist might feel that the choice between the Jaguar line and the Donne line will be decided in terms of strategic exigency.

Reference-group theory indicates that social affiliation or disaffiliation normally carries with it specific cognitive commitments. One joins one group and thereby 'knows' that the world is such-and-such. One leaves this group for another and now 'knows' that one must have been mistaken. Every group to which one refers oneself occupies a vantage point on the universe. Every role has a world view dangling from its end. In choosing specific people one chooses a specific world to live in. If the sociology of knowledge gives us a broad view of the social construction of reality, reference-group theory shows us the many little workshops in which cliques of universe-builders hammer out their models of the cosmos. The socio-psychological dynamics that underlies this process is presumably the same we have already looked at in connexion with role theory – the primeval human urge to be accepted, to belong, to live in a world with others.

Some of the experiments conducted by social psychologists on the way in which group opinion affects even the perception of physical objects give us an awareness of the irresistible force of this urge. An individual confronted with an object that is, say, thirty inches in length will progressively modify his initially correct estimate if placed in an experimental group all the members of which keep repeating that they are quite sure about the actual length being ten inches or so. It should not surprise us, consequently, that group opinions in political, ethical or aesthetic matters should exercise even greater force, since the individual thus pressured cannot have desperate recourse to a political, ethical or aesthetic tape measure. And if he should try, of course, the group will deny that his measure is a measure in the first place. One group's measure of validity is another's gauge of benightedness. Criteria of canonization and anathematization are interchangeable. One chooses one's gods by choosing one's playmates.

In this chapter we have singled out some strands of sociological thought that present us with a picture of society existing within man, adding to our previous perspective on man within society. At this point, our picture of society as a great prison no longer seems satisfactory, unless we add to it the detail of groups of prisoners busily keeping its walls intact. Our imprisonment in society now appears as something affected as much from within ourselves as by the operation of external forces. A more adequate representation of social reality now would be the puppet theatre, with the curtain rising on the little puppets jumping about on the ends of their invisible strings, cheerfully acting out the little parts that have been assigned to them in the tragi-comedy to be enacted. The analogy, however, does not go far enough. The Pierrot of the puppet theatre has neither will nor consciousness. But the Pierrot of the social stage wants nothing more than the fate awaiting him in the scenario – and he has a whole system of philosophy to prove it.

The key term used by sociologists to refer to the phenomena discussed in this chapter is that of internalization. What happens in socialization is that the social world is internalized within the child. The same process, though perhaps weaker in quality, occurs every time the adult is initiated into a new social context or a new social group. Society, then, is not only something 'out there', in the Durkheimian sense, but it is also 'in here', part of our innermost being. Only an understanding of internalization makes sense of the incredible fact that most external controls work most of the time for most of the people in a society. Society not only controls our movements, but shapes our identity, our thought and our emotions. The structures of society become the structures of our own consciousness. Society does not stop at the surface of our skins. Society penetrates us as much as it envelops us. Our bondage to society is not so much established by conquest as by collusion. Sometimes, indeed, we are crushed into submission. Much more frequently we are entrapped by our own social nature.

The walls of our imprisonment were there before we appeared on the scene, but they are ever rebuilt by ourselves. We are betrayed into captivity with our own cooperation.

6

SOCIETY AS DRAMA

IF the attempt at communication in the preceding two chapters has been successful, the reader may now have a sensation that could perhaps be described as sociological claustrophobia. He can be conceded a certain moral right to demand some relief from this from the writer, in the way of an affirmation of human freedom in the face of the various social determinants. Such an affirmation, however, poses *a priori* difficulties within the framework of a sociological argument. It will be necessary to look at these difficulties briefly before we proceed.

Freedom is not empirically available. More precisely, while freedom may be experienced by us as a certainty along with other empirical certainties, it is not open to demonstration by any scientific methods. If we wish to follow Kant, freedom is also not available rationally, that is, cannot be demonstrated by philosophical methods based on the operations of pure reason. Remaining here with the question of empirical availability, the elusiveness of freedom with regard to scientific comprehension does not lie so much in the unspeakable mysteriousness of the phenomenon (after all, freedom may be mysterious, but the mystery is encountered every day) as in the strictly limited scope of scientific methods. An empirical science must operate within certain assumptions, one of which is that of universal causality. Every object of scientific scrutiny is presumed to have an anterior cause. An object, or an event, that *is* its own cause lies outside the scientific universe of discourse. Yet freedom has precisely that character. For this reason, no amount of scientific search will ever uncover a phenomenon that can be designated as free. Whatever may appear as free within the subjective consciousness of an individual will find its place in the scientific scheme as a link in some chain of causation.

Freedom and causality are not logically contradictory terms. However, they are terms that belong to disparate frames of reference. It is, therefore, idle to expect that scientific methods will be able to uncover freedom by some procedure of elimination, piling up causes on causes, until one arrives at a residual phenomenon that does not seem to have a cause and can be proclaimed as being free. Freedom is not that which is uncaused. Similarly, one cannot arrive at freedom by looking at instances where scientific prediction falls down. Freedom is not unpredictability. As Weber has shown, if this were the case, the madman would be the freest human being. The individual who is conscious of his own freedom does not stand outside the world of causality, but rather perceives his own volition as a very special category of cause, different from the other causes that he must reckon with. This difference, however, is not subject to scientific demonstration.

An analogy may be helpful here. Just as freedom and causality are not contradictory but rather disparate terms, so are utility and beauty. The two do not logically exclude each other. But one cannot establish the reality of the one by demonstrating the reality of the other. It is possible to take a specific object, say a piece of furniture, and show conclusively that it has a certain utility for human living – to sit on, eat on, sleep in, or what-have-you. However, no matter what utility one can prove, one will get no closer to the question of whether that chair, table or bed is beautiful. In other words, the utilitarian and the aesthetic universes of discourse are strictly incommensurable.

In terms of social-scientific method, one is faced with a way of thinking that assumes *a priori* that the human world is a causally closed system. The method would not be scientific if it thought otherwise. Freedom as a special kind of cause is excluded from this system *a priori*. In terms of social phenomena, the social scientist must assume an infinite regress of causes, none of them holding a privileged ontological status. If he cannot explain a phenomenon causally by one set of sociological categories, he will try another one. If political causes do not seem satisfactory, he will try economic ones. And if the entire conceptual apparatus of sociology seems inadequate to explain a given phenomenon, he may switch to a different apparatus, such as the psychological or the biological one. But in doing so, he will still move within the scientific cosmos – that is, he will discover new orders of causes, but he will not encounter freedom. There is no way of perceiving freedom, either in oneself or in another human being, except through a subjective inner certainty that dissolves as soon as it is attacked with the tools of scientific analysis.

Nothing is farther from the intentions of this writer than to come out now with a statement of allegiance to that positivistic creed, still fashionable among some American social scientists, that believes in only those fragments of reality that can be dealt with scientifically. Such positivism results almost invariably in one form or another of intellectual barbarism, as has been demonstrated admirably in the recent history of behaviouristic psychology in this country. Nevertheless, one must keep a kosher kitchen if one's intellectual nourishment is not to become hopelessly polluted – that is, one must not pour the milk of subjective insight over the meat of scientific interpretation. Such segregation does not mean that one cannot relish both forms of sustenance, only that one cannot do so in a single dish.

It follows that if our argument wanted to remain rigidly within the sociological frame of reference, which is a scientific one, we could not speak about freedom at all. We would then have to leave the reader to his own devices in getting out of his claustrophobic corner. Since these lines, fortunately, do not appear in a sociological journal and are not to be recited at a ceremonial gathering of the profession, there is no need to be as ascetic as all that. Instead, we shall follow two courses. First, still remaining within the model of human existence provided by sociological perspective itself, we shall try to show that the controls, both external and internal, may not be as infallible as they were made to appear so far. Secondly, we shall step outside the narrowly scientific frame of reference and *postulate* the reality of freedom, after which we shall try to see what the sociological model looks like from the vantage point of this postulation. In the first course, we shall give some further touches to our sociological perspective. In the second, we shall seek to obtain some human perspective *on* sociological perspective.

Let us return to the point in our argument, at the end of the last chapter, in which we maintained that our own cooperation is needed to bring us into social captivity. What is the nature of this cooperation? One possibility of answering this question is to take up once more Thomas's concept of the definition of the situation. We can then argue that, whatever the external and internal pressures of society may be, in most cases we ourselves must be at least co-definers of the social situation in question. That is, whatever the prehistory of this may be, we ourselves are called to an act of collaboration in the maintenance of the particular definition. However, another possibility of getting at the above question is to switch to another system of sociological conceptualization, namely that of Weber. We contend that a Weberian approach at this point will serve as a helpful balance to the Durkheimian angle on social existence.

Talcott Parsons has compared Weberian sociology to other approaches by calling it 'voluntaristic'. Although Weber's conception of scientific methodology was far too Kantian to allow for the introduction of the idea of freedom into his system, Parsons's term is apt in distinguishing the Weberian emphasis on the intentionality of social action as against the Durkheimian lack of interest in this dimension. As we have seen, Durkheim stresses the externality, objectivity, 'thing'-like character of social reality (one is almost tempted here to use the scholastic term of 'quiddity'). Against this, Weber always emphasizes the subjective meanings, intentions and interpretations brought into any social situation by the actors participating in it. Weber, of course, also points out that what eventually happens in society may be very different from what these actors meant or intended. But he asserts that this entire subjective dimension must be taken into consideration for an adequate sociological understanding. (*Verstehen* is the technical term used to denote the latter, a term that has been taken over into English sociological parlance.) That is, sociological understanding involves the interpretation of meanings present in society.

In this view, each social situation is sustained by the fabric of meanings that are brought into it by the several participants. It is clear, of course, that in a situation whose meaning is strongly established by tradition and common consent a single individual cannot accomplish very much by proffering a deviant definition. At the very least, however, he can bring about his alienation from the situation. The possibility of marginal existence in society is already an indication that the commonly agreed-upon meanings are not omnipotent in their capacity to coerce. But more interesting are those cases where individuals succeed in capturing enough of a following to make their deviant interpretations of the world stick, at least within the circle of this following.

This possibility of breaking through the 'world-taken-for-granted' of a society is developed in Weber's theory of charisma. The term, derived from the New Testament (where, however, it is used in a very different sense), denotes social authority that is not based on tradition or legality but rather on the extraordinary impact of an individual leader. The religious prophet, who defies the established order of things in the name of an absolute authority given to him by divine command, is the prototype of the charismatic leader. One can think here of historical figures such as the Buddha, Jesus or Muhammad. Charisma, however, can also appear in the profane areas of life, especially the political one. One can here think of such personages as Caesar or Napoleon. The paradigmatic form of such charismatic authority setting itself up against the established order can be found in Jesus's reiterated assertions that 'you have heard it said . . . but I say to you'. In this 'but' lies a claim rightfully to supersede whatever was regarded as binding before. Typically, then, charisma constitutes a tremendously passionate challenge to the power of predefinition. It substitutes new meanings for old and radically redefines the assumptions of human existence.

Charisma is not to be understood as some sort of miracle that occurs without reference to what has happened before or to the social context of its appearance. Nothing in history is free of ties with the past. Also, as Weber's theory of charisma has developed in great detail, the extraordinary passion of a charismatic movement only rarely survives for longer than one generation. Invariably charisma becomes what Weber called 'routinized', that is, becomes reintegrated into the structures of society in much less radical forms. Prophets are followed by popes, revolutionaries by administrators. When the great cataclysm of religious or political revolution is over and men have settled down to live under what was considered a new order, it invariably turns out that the changes have not been quite as total as it first appeared. Economic interests and political ambitions take over at the point where insurrectionary fervour has begun to cool. The old habits reassert themselves and the order created by the charismatic revolution begins to acquire disturbing similarities with the *ancien régime* that it overthrew with so much violence. Depending on one's values, this fact may sadden or comfort one. What interests us, however, is not the long-range weakness of rebellion in history, but its possibility in the first place.

It is noteworthy in this connexion that Weber regarded charisma as one of the principal moving forces in history, despite his clear insight into the fact that charisma is always a very short-lived phenomenon. But however much the old patterns may reappear in the course of the 'routinization' of charisma, the world is never quite the same again. Even though the change has been much less than the revolutionaries hoped or expected, change there has been none the less. Sometimes only the passage of much time shows just how deep the change has gone. This is why almost all attempts at total counter-revolution fail in history, as such undertakings as the Council of Trent or the Congress of Vienna illustrate. The lesson to be derived from this for our sociological perspective is simple, almost platitudinous, but none the less significant for a more balanced picture: it is possible to challenge effectively the Leviathan of predefinition. Or to put the same thing negatively, in terms of our previous discussion: it is possible to withold our co-operation with history.

Part of the inexorable impression conveyed by the Durkheimian and related views of society comes from their not paying sufficient attention to the historical process itself. No social structure, however massive it may appear in the present, existed in this massivity from the dawn of time. Somewhere along the line each one of its salient features was concocted by human beings, whether they were charismatic visionaries, clever crooks, conquering heroes or just individuals in positions of power who hit on what seemed to them a better way of running the show. Since all social systems were created by men, it follows that men can also change them. Indeed, one of the limitations of the afore-mentioned views of society (which, to emphasize this again, give us a valid perspective on social reality) is that it is difficult to account for the change within their frame of reference. This is where the historical orientation of the Weberian approach redresses the balance.

The Durkheimian and Weberian ways of looking at society are not logically contradictory. They are only anti-thetical since they focus on different aspects of social reality. It is quite correct to say that society is objective fact, coercing and even creating us. But it is also correct to say that our own meaningful acts help to support the edifice of society and may on occasion help to change it. Indeed, the two statements contain between them the paradox of social existence: that society defines us, but is in turn defined by us. This paradox is what we have alluded to before in terms of our collusion and collaboration with society. As soon as we view society in this way, however, it appears very much more fragile than it did from the other vantage point. We need the recognition of society to be human, to have an image of ourselves, to have an identity. But society needs the recognition of many like us in order to exist at all. In other words, it is not only ourselves but society that exists by virtue of definition. It will depend on our social location as to whether our refusal to recognize a particular social reality will have much of an effect. It does not help the slave much to refuse to recognize his enslavement. It is a different story when one of the masters does so. However, systems of slavery have always reacted violently to such a challenge even from their humblest victims. It would seem, then, that just as there is no total power in society, there is also no total impotence. The masters in society recognize this fact and apply their controls accordingly.

It follows that the control systems are in constant need of confirmation and reconfirmation by those they are meant to control. It is possible to withold such confirmation in a number of ways. Each one constitutes a threat to society as defined officially. The possibilities to take into consideration here are those of transformation, detachment and manipulation.

Our reference to charisma has already indicated in what way the transformation of social definitions may occur. Charisma, of course, is not the only factor that can induce change in society. Any process of social change, however, is connected with new definitions of reality. Any such redefinition means that someone begins to act contrary to expectations directed towards him in line with the old definition. The master expects a bow from his slave and instead gets a fist in his face. It will depend, of course, on how frequent such incidents are whether we speak of individual 'deviance' or social 'disorganization', to use common sociological terms. When an individual refuses to recognize the social definition of economic rights, we will be faced with a phenomenon of crime, namely with those acts of deviance that are listed in the F B I statistics as 'crimes against property'. But when masses of individuals, under political leadership, engage in the same refusal, we confront a revolution (be it in the form of the establishment of a socialist order or, more mildly, in a radical new tax system). The sociological differences between individual deviance, such as crime, and the wholesale dis- and re-organization of an entire social system, such as revolution, are obvious. Both, however, are significant in terms of our argument, in showing the possibility of resistance to the external and (of necessity) also the internal controls. In fact, when we look at revolutions, we find that the outward acts against the old order are invariably preceded by the disintegration of inward allegiance and loyalties. The images of kings topple before their thrones do. As Albert Salomon has shown, this destruction of the peoples' conception of their rulers can be illustrated by the Affair of the Queen's Necklace before the French Revolution and the Rasputin case before the Russian. The ongoing insurrection of Southern Negroes against the segregation system in our own time was similarly preceded by a long process in which the old definitions of their role were discredited in the nation at large and destroyed in their own minds (a process, by the way, in which social scientists, including white Southern ones, played a not insignificant part). In other words, long before social systems are brought down in violence, they are deprived of their ideological sustenance by contempt. Non-recognition and counter-definition of social norms are always potentially revolutionary.

However, we can look at much more routine cases in which particular social situations can be transformed or at least sabotaged by a refusal to accept their previous definitions. If we may make a rather unscholarly reference here, we would point to the opus of the English humourist Stephen Potter as an excellent guide to the subtle art of social sabotage. What Potter calls the 'ploy' is precisely the technique of redefining a situation contrary to general expectations – and doing so in such a way that the other participants in the situation are caught off guard and find themselves helpless to counter-attack. The patient who prearranges phone calls in such a way that he converts his doctor's consultation room into a business office, the American tourist in England who lectures his English host on the antiquities of London, the non-churchgoing house-guest who manages thoroughly to upset his churchgoing hosts on Sunday morning by alluding to his own darkly esoteric religious preference that would not possibly permit him to join them – all these are instances of what could be called successful micro-sociological sabotage, picayune compared to the Promethean *bouleversements* of the great revolutionary, but none the less revealing of the innate precariousness of the social fabric. If his moral prejudices allow, the reader can readily test the validity of the Potterite technique of sociological demolition (which might well be called, with due apologies to Madison Avenue, the engineering of dissent). Let him pretend to be a tolerant but firm abstainer at a New York cocktail party, or an initiate of some mystic cult at a Methodist church picnic, or a psychoanalyst at a business-men's luncheon – in each case, he is quite likely to find that the introduction of a dramatic character that does not fit into the scenario of the particular play seriously threatens the role-playing of those who do fit. Experiences such as these may lead to a sudden reversal in one's view of society – from an awe-inspiring vision of an edifice made of massive granite to the picture of a toy-house precariously put together with *papier mâché*. While such metamorphosis may be disturbing to people who have hitherto had great confidence in the stability and rightness of society, it can also have a very liberating effect on those more inclined to look upon the latter as a giant sitting on top of them, and not necessarily a friendly giant at that. It is reassuring to discover that the giant is afflicted with a nervous tic.

If one cannot transform or sabotage society, one can withdraw from it inwardly. Detachment has been a method of resistance to social controls at least since Lao-tzu and was made into a theory of resistance by the Stoics. The person who retires from the social stage into religious, intel-lectual or artistic domains of his own making still, of course, carries into this self-imposed exile the language, identity and store of knowledge that he initially achieved at the hands of society. Nevertheless it is possible, though fre-quently at considerable psychological cost, to build for oneself a castle of the mind in which the day-to-day expectations of society can be almost completely ignored. And as one does this, the intellectual character of this castle is more and more shaped by oneself rather than by the ideologies of the surrounding social system. If one finds others to join one in such an enterprise, one can in a very real sense create a counter-society whose relations with the other, the 'legitimate' society can be reduced to a diplomatic minimum. Incidentally, in that case the psychological burden of such detachment can be greatly minimized.

Such counter-societies, constructed on the basis of deviant and detached definitions, exist in the form of sects, cults, 'inner circles' or other groups that sociologists call sub-cultures. If we want to emphasize the normative and cognitive separateness of such groups, the term sub-world may be an apter one. A sub-world exists as an island of deviant meanings within the sea of its society, to adapt the phrase that Carl Mayer used eloquently to describe the social character of religious sectarianism. The individual who enters such a sub-world from the outside is made to feel very strongly that he is entering an entirely different universe of discourse. Eccentric religiosity, subversive politics, unconventional sexuality, illegal pleasures – any of these are capable of creating a sub-world carefully shielded from the effect of both the physical and the ideological controls of the larger society. Thus a modern American city may contain, well hidden from public view, its subterranean worlds of theosophists, Trotskyists, homosexuals or drug addicts, speaking their own language and in its terms building a universe infinitely far removed in meaning from the world of their fellow citizens. Indeed, the anonymity and freedom of movement of modern urban life greatly facilitate the building of such underworlds.

However, it is important to emphasize that less rebellious constructions of the mind can also liberate the individual to a considerable extent from the definitory system of his society. A man who passionately devotes his life to the study of pure mathematics, theoretical physics, Assyriology or Zoroastrianism can afford to pay a minimum of attention to routine social demands, as long as he can somehow manage to survive economically in the pursuit of his interests. And, what is more important, the directions of thinking that these universes of discourse will naturally lead him to will have a very high degree of autonomy indeed *vis-à-vis* the routine intellectual patterns that constitute the world view of the man's society. One may recall here the toast delivered at a gathering of mathematicians: 'To pure mathematics – and may it never be of any use to anybody!' Unlike some of the examples mentioned earlier, this kind of sub-world does not arise out of rebellion against society as such, but it leads all the same to an autonomous intellectual universe within which an individual can exist with almost Olympic detachment. Put differently, it is possible for men, alone or in groups, to construct their own worlds and on this basis to detach themselves from the world into which they were originally socialized.

The discussion of the art of 'ploying' has already brought us close to the third major way of escaping the tyranny of society, that of manipulation. Here the individual does not try to transform the social structures nor does he detach himself from them. Rather he makes deliberate use of them in ways unforeseen by their legitimate guardians, cutting a path through the social jungle in accordance with his own purposes. Erving Goffman, in his analysis of the world of 'inmates' (be it of mental hospitals or prisons or other coercive institutions), has given us vivid examples of how it is possible to 'work the system', that is, to utilize it in ways not provided for in the official operating procedures. The convict who works in the prison laundry and uses its machinery to wash his own socks, the patient who get access to the staff communications system to transmit personal messages, the soldier who manages to transport his girl friends in military vehicles – all these are 'working the system', thereby proclaiming their own relative independence of its tyrannical demands. It would be rash to dismiss such manipulations too quickly as pathetic and ineffective efforts at rebellion. There have been instructive cases in which motor-pool sergeants successfully ran call-girl rings and hospital patients used the official message centre as a bookie joint, such operations going on in subterranean fashion for long periods of time. And industrial sociology is full of examples of how workers can employ the official organization of a factory for purposes deviant from and sometimes contradictory to the intentions of management.

The ingenuity human beings are capable of in circumventing and subverting even the most elaborate control system is a refreshing antidote to sociologistic depression. It is as relief from social determinism that we would explain the sympathy that we frequently feel for the swindler, the imposter or the charlatan (as long, at any rate, as it is not ourselves who are being swindled). These figures symbolize a social Machiavellianism that understands society thoroughly and then, untrammelled by illusions, finds a way of manipulating society for its own ends. In literature there are characters such as André Gide's Lafcadio or Thomas Mann's Felix Krull that illustrate this fascination. In real life we could point to a man like Ferdinand Waldo Demara, Jr., who bamboozled a long line of eminent specialists in various fields into accepting him as a colleague, successfully impersonating such respected social identities as college professor, military officer, penologist and even surgeon. Inevitably, in watching the swindler take on various roles of respectable society, we are pushed towards the uncomfortable impression that those who hold these roles 'legitimately' may have attained their status by procedures not so drastically different from the ones employed by him. And if one knows the bamboozling, bunkum and (to use Potter's term) 'one-upmanship' that go into, say, a professorial career one may even come dangerously close to the conclusion that society is a swindle to begin with. In one way or another, we are all imposters. The ignoramus impersonates erudition, the crook honesty, the sceptic conviction – and any normal university could not exist without the first confidence trick, no business organization without the second and no church without the third.

Another concept elaborated by Goffman is helpful in this connexion – the one he calls 'role distance'. By this Goffman means the playing of a role tongue-in-cheek, without really meaning it and with an ulterior purpose. Every strongly coercive situation will produce this phenomenon. The 'native' underling plays up to the *pukka sahib* in the expected way while planning the day on which all white throats will be cut. The Negro domestic plays the role of self-depreciating clown, and the enlisted man that of spick-and-span military fanatic, both with hindthoughts that are diametrically contrary to the mythology within which their roles have a meaning they inwardly reject. As Goffman points out, this kind of duplicity is the only way by which human dignity can be maintained within the self-awareness of people in such situations. But Goffman's concept could be applied more widely to all cases where a role is played deliberately without inner identification, in other words, where the actor has established an inner distance between his consciousness and his role-playing. Such cases are of paramount importance for sociological perspective because they depart from the normal pattern. This, as we have been at pains to point out, is that roles are played without reflection, in immediate and almost automatic response to the expectations of the situation. Here this fog of unconsciousness is suddenly dispelled. In many instances this may not affect the visible course of events, yet it constitutes a qualitatively different form of existence in society. 'Role distance' marks the point at which the marionette clown becomes Bajaccio – the puppet theatre is transformed into a living stage. Of course, there is still a script, a stage management and a repertoire that includes one's own role. But one is now playing the part in question with full consciousness. As soon as this happens, there is the ominous possibility that Bajaccio may jump out of his role and start playing the tragic hero – or that Hamlet may begin to do somersaults and sing dirty ditties. Let us repeat our previous assertion that all revolutions begin in transformations of consciousness.

A useful concept to introduce in this connexion is that of 'ecstasy'. By this we refer not to some abnormal heightening of consciousness in a mystic sense, but rather, quite literally, to the act of standing or stepping outside (literally, *ekstasis*) the taken-for-granted routines of society. In our discussion of 'alternation' we have already touched upon a very important form of 'ecstasy' in our sense, namely, the one that takes place when an individual is enabled to jump from world to world in his social existence. However, even without such an exchange of universes it is possible to achieve distance and detachment *vis-à-vis* one's own world. As soon as a given role is played without inner commitment, deliberately and deceptively, the actor is in an ecstatic state with regard to his 'world-taken-for-granted'. What others regard as fate, he looks upon as a set of factors to reckon with in his operations. What others assume to be essential identity, he handles as a convenient disguise. In other words, 'ecstasy' transforms one's awareness of society in such a way that *givenness* becomes *possibility*. While this begins as a state of consciousness, it should be evident that sooner or later there are bound to be significant consequences in terms of action. From the point of view of the official guardians of order, it is dangerous to have too many individuals around playing the social game with inner reservations.

The consideration of 'role distance' and 'ecstasy' as possible elements of social existence raises an interesting sociology-of-knowledge question, namely, whether there are social contexts or groups that particularly facilitate such consciousness. Karl Mannheim, who greatly favoured such a development on ethical and political grounds (a position that some might want to debate), spent a good deal of time looking for its possible social ground. His view of the 'freely suspended intelligentsia' (that is, of a stratum of intellectuals with minimal involvement in the vested interests of society) as the best carriers of this sort of liberated consciousness may be disputed. At the same time, there can be little doubt that certain kinds of intellectual training and activity are capable of leading to 'ecstasy', as we indicated in our discussion of the forms of detachment.

Other tentative generalizations can be made. 'Ecstasy' is more likely to take place in urban than in rural cultures (*vide* the classic role of cities as places of political freedom and liberality in thought), among groups that are marginal to society than among those at its centre (*vide* the historic relationship of European Jews to various liberating intellectual movements – or, in a very different way, take the example of the itinerant Bulgarian journeymen carrying the Manichaean heresy all the way across Europe into Provence), as it is also more likely in groups that are insecure in their social position than among those that are secure (*vide* the production of debunking ideologies among rising classes that have to fight against an established order, the rising French bourgeoisie in the seventeenth and eighteenth centuries providing us with a prime example). Such social location of the phenomenon reminds us once more that not even total rebellion takes place in a social vacuum without predefinitions. Even nihilism is predefined in terms of the structures it is driven to negate – before one can have atheism, for instance, there must be an idea of God. In other words, every liberation from social roles takes place within limits that are social themselves. Nevertheless, our consideration of the various forms of 'ecstasy' has taken us some way from the deterministic corner into which our previous argument had chased us.

We thus arrive at a third picture of society, after those of the prison and the puppet theatre, namely that of society as a stage populated with living actors. This third picture does not obliterate the previous two, but it is more adequate in terms of the additional social phenomena we have considered. That is, the dramatic model of society at which we have arrived now does not deny that the actors on the stage are constrained by all the external controls set up by the impresario and the internal ones of the role itself. All the same, they have options – of playing their parts enthusiastically or sullenly, of playing with inner conviction or with 'distance', and, sometimes, of refusing to play at all. Looking at society through the medium of this dramatic model greatly changes our general sociological perspective. Social reality now seems to be precariously perched on the cooperation of many individual actors – or perhaps a better simile would be that of acrobats engaged in perilous balancing acts, holding up between them the swaying structure of the social world.

Stage, theatre, circus and even carnival – here we have the imagery of our dramatic model, with a conception of society as precarious, uncertain, often unpredictable. The institutions of society, while they do in fact constrain and coerce us, appear at the same time as dramatic conventions, even fictions. They have been invented by past impresarios, and future ones may cast them back into the nothingness whence they emerged. Acting out the social drama we keep pretending that these precarious conventions are eternal verities. We act *as if* there were no other way of being a man, a political subject, a religious devotee or one who exercises a certain profession – yet at times the thought passes through the minds of even the dimmest among us that we could do very, very different things. If social reality is dramatically created, it must also be dramatically malleable. In this way, the dramatic model opens up a passage out of the rigid determinism into which sociological thought originally led us.

Before we leave behind us our narrower sociological argument we would like to point to a classical contribution that is very relevant to the points just made – the theory of sociability of the German sociologist Georg Simmel, a contemporary of Weber's whose approach to sociology differed considerably from the latter's. Simmel argued that sociability (in the usual meaning of this word) is the play-form of social interaction. At a party people 'play society', that is, they engage in many forms of social interaction, but without their usual sting of seriousness. Sociability changes serious communication to noncommittal conversation, *eros* to coquetry, ethics to manners, aesthetics to taste. As Simmel shows, the world of sociability is a precarious and artificial creation that can be shattered at any moment by someone who refuses to play the game. The man who engages in passionate debate at a party spoils the game, as does the one who carries flirtation to the point of open seduction (a party is *not* an orgy) or the one who openly promotes business interests under the guise of harmless chitchat (party conversation must at least pretend to be disinterested). Those who participate in a situation of pure sociability temporarily leave behind their 'serious' identities and move into a transitory world of make-believe, which consists among other things of the playful pretence that those concerned have been freed from the weights of position, property and passions normally attached to them. Anyone who brings in the gravity (in both senses of the word) of 'serious' outside interests immediately shatters this fragile artifice of make-believe. This, incidentally, is why pure sociability is rarely possible except among social equals, since otherwise the pretence is too strenuous to maintain – as every office party shows painfully.

We are not particularly interested in the phenomenon of sociability for its own sake, but we can now relate what Simmel maintains about it to our earlier consideration of Mead's notion that social roles are learned through play. We contend that sociability could not exist at all as the artifice it is if society at large did not have a similarly artificial character. In other words, sociability is a special case of 'playing society', more consciously fictitious, less tied up with the urgent ambitions of one's career – but yet of one piece with a much larger social fabric that one can also play with. It is precisely through such play, as we have seen, that the child learns to take on his 'serious' roles. In sociability we return for some moments to the masquerading of childhood – hence perhaps the pleasure of it.

But it is assuming too much to think that the masks of the 'serious' world are terribly different from those of this world of play. One plays the masterful *raconteur* at the party and the man of firm will at the office. Party tact has a way of being translated into political finesse, shrewdness in business into the adroit handling of etiquette for purposes of sociability. Or, if you like, there is a nexus between 'social graces' and social skills in general. In this fact lies the sociological justification of the 'social' training of diplomats as well as of debutantes. By 'playing society' one learns how to be a social actor anywhere. And this is possible only because society as a whole has the character of a play. As the Dutch historian Johan Huizinga has brilliantly shown in his book *Homo ludens*, it is impossible to grasp human culture at all unless we look at it *sub specie ludi* – under the aspect of play and playfulness.

With these thoughts we have come to the very limits of what it is still possible to say within a social-scientific frame of reference. Within the latter, we cannot go any farther in lifting from the reader the deterministic burden of our earlier argumentation. Compared with this argumentation, what has been said in the present chapter so far may appear rather weak and less than conclusive. This is unavoidable. To repeat ourselves, it is impossible *a priori* to come upon freedom in its full sense by scientific means or within a scientific universe of discourse. The closest we have been able to come is to show, in certain situations, a certain freedom *from* social controls. We cannot possibly discover freedom *to* act socially by scientific means. Even if we should find holes in the order of causality that can be established sociologically, the psychologist, the biologist or some other dealer in causations will step in and stuff up our hole with materials spun from *his* cloth of determinism. But since we have made no promises in this book to limit ourselves ascetically to scientific logic, we are now ready to approach social existence from a very different direction. We have not been able to get at freedom sociologically, and we realize that we never can. So be it. Let us see now how we can look at our sociological model itself from a different vantage point.

As we remarked before, only an intellectual barbarian is likely to maintain that reality is only that which can be grasped by scientific methods. Since, hopefully, we have tried to stay out of this category, our sociologizing has been carried on in the foreground of another view of human existence that is not itself sociological or even scientific. Nor is this view particularly eccentric, but rather the common (if very differently elaborated) anthropology of those who credit man with the capacity for freedom. Obviously a philosophical discussion of such an anthropology would utterly break the framework of this book and would, for that matter, lie beyond the competence of its writer. But while no attempt will be made here to provide a philosophical introduction to the question of human freedom, it is necessary to our argument that at least some indications be given of how it is possible to think sociologically without abandoning this notion of freedom, and, more than that, in what way a view of man that includes the idea of freedom may take cognizance of social dimension. We contend that here is an important area of dialogue between philosophy and the social sciences that still contains vast tracts of virgin territory. We point to the work of Alfred Schuetz and to the contemporary efforts of Maurice Natanson as indicating the direction in which this dialogue could move. Our own remarks in the following pages will, of necessity, be exceedingly sketchy. But it is hoped that they will suffice to indicate to the reader that sociological thought need not necessarily end in a positivistic swamp.

We shall now begin with the postulate that men are free and from this new starting point return to the same problem of social existence. In doing this, we shall find helpful some concepts developed by existentialist philosophers (though we shall use these without any doctrinaire intentions). Herewith the reader is invited to undertake an epistemological *salto mortale* – and this behind him, to return to the matter at hand.

Let us retrace our steps to the point where we looked at Gehlen's theory of institutions. The latter, we will recall, are interpreted in this theory as channelling human conduct very much along the lines that instincts channel the behaviour of animals. When we considered this theory, we made the remark that there is, however, one crucial difference between the two kinds of channelling: the animal, if it reflected on the matter of following its instincts, would say,

'I have no choice'. Men, explaining why they obey their institutional imperatives, say the same. The difference is that the animal would be saying the truth; the men deceiving themselves. Why? Because, in fact, they *can* say 'no' to society, and often have done so. There may be very unpleasant consequences if they take this course. They may not even think about it as a possibility, because they take their own obedience for granted. Their institutional character may be the only identity they can imagine having, with the alternative seeming to them as a jump into madness. This does not change the fact that the statement 'I must' is a deceptive one in almost every social situation.

From our new vantage point, within an anthropological frame of reference that recognizes man as free, we can usefully apply to this problem what Jean-Paul Sartre has called 'bad faith'. To put it very simply, 'bad faith' is to pretend something is necessary that in fact is voluntary. 'Bad faith' is thus a flight from freedom, a dishonest evasion of the 'agony of choice'. 'Bad faith' expresses itself in innumerable human situations from the most commonplace to the most catastrophic. The waiter shuffling through his appointed rounds in a café is in 'bad faith' insofar as he pretends to himself that the waiter role constitutes his real existence, that, if only for the hours he is hired, he *is* the waiter. The woman who lets her body be seduced step by step while continuing to carry on an innocent conversation is in 'bad faith', insofar as she pretends that what is happening to her body is not under her control. The terrorist who kills and excuses himself by saying that he had no choice because the party ordered him to kill is in 'bad faith', because he pretends that his existence is necessarily linked with the party, while in fact this linkage is the consequence of his own choice. It can easily be seen that 'bad faith' covers society like a film of lies. The very possibility of 'bad faith', however, shows us the reality of freedom.

164

Man can be in 'bad faith' only because he is free and does not wish to face his freedom. 'Bad faith' is the shadow of human liberty. Its attempt to escape that liberty is doomed to defeat. For, as Sartre has famously put it, we are 'condemned to freedom'.

If we apply this concept to our sociological perspective, we will suddenly be faced with a startling conclusion. The complex of roles within which we exist in society now appears to us as an immense apparatus of 'bad faith'. Each role carries with it the possibility of 'bad faith'. Every man who says 'I have no choice' in referring to what his social role demands of him is engaged in 'bad faith'. Now, we can easily imagine circumstances in which this confession will be true to the extent that there is no choice *within that particular role*. Nevertheless, the individual has the choice of stepping outside the role. It is true that, given certain circumstances, a businessman has 'no choice' but brutally to destroy a competitor, unless he is to go bankrupt himself, but it is he who chooses brutality over bankruptcy. It is true that a man has 'no choice' but to betray a homosexual attachment if he is to retain his position in respectable society, but he is the one making the choice between respectability and loyalty to that attachment. It is true that in some cases a judge has 'no choice' but to sentence a man to death, but in doing so he chooses to remain a judge, an occupation chosen by him in the knowledge that it might lead to this, and he chooses not to resign instead when faced with the prospect of this duty. Men are responsible for their actions. They are in 'bad faith' when they attribute to iron necessity what they themselves are choosing to do. Even the law itself, that master fortress of 'bad faith', has begun to take cognizance of this fact in its dealings with Nazi war criminals.

Sartre has given us a masterful vista of the operation of 'bad faith' at its most malevolent in his portrayal of the anti-Semite as a human type. The anti-Semite is the man who frantically identifies himself with mythological entities ('nation', 'race', '*Volk*') and in doing so seeks to divest himself of the knowledge of his own freedom. Anti-Semitism (or, we might add, any other form of racism or fanatical nationalism) is 'bad faith' *par excellence* because it identifies men in their human totality with their social character. Humanity itself becomes a facticity devoid of freedom. One then loves, hates and kills within a mythological world in which all men *are* their social designations, as the SS man *is* what his insignia says and the Jew *is* the symbol of despicability sewn on his concentration-camp uniform.

'Bad faith' in this form of ultimate malignancy, however, is not limited to the Kafkaesque world of Nazism and its totalitarian analogies. It exists in our own society in identical patterns of self-deception. It is only as one long series of acts of 'bad faith' that capital punishment continues to exist in allegedly humane societies. Our torturers, just like the Nazi ones, present themselves as conscientious public servants, with an impeccable if mediocre private morality, who reluctantly overcome their weakness in order to do their duty.

We will not at this point go into the ethical implications of such 'bad faith'. We shall do so briefly in the digression that follows this chapter. We would rather return here to the startling view of society that we have reached as a result of these considerations. Since society exists as a network of social roles, each one of which can become a chronic or a momentary alibi from taking responsibility for its bearer, we can say that deception and self-deception are at the very heart of social reality. Nor is this an accidental quality that could somehow be eradicated by some moral reformation or other. The deception inherent in social structures is a functional imperative. Society can maintain itself only if its fictions (its 'as if' character, to use Hans Vaihinger's term) are accorded ontological status by at least some of its members some of the time – or, let us say, society as we have so far known it in human history.

Society provides for the individual a gigantic mechanism by which he can hide from himself his own freedom. Yet this character of society as an immense conspiracy in 'bad faith' is, just as in the case of the individual, but an expression of the possibility of freedom that exists by virtue of society. We are social beings and our existence is bound to specific social locations. The same social situations that can become traps of 'bad faith' can also be occasions for freedom. Every social role can be played knowingly or blindly. And insofar as it is played knowingly, it can become a vehicle of our own decisions. Every social institution can be an alibi, an instrument of alienation from our freedom. But at least some institutions can become protective shields for the actions of free men. In this way, an understanding of 'bad faith' does not necessarily lead us to a view of society as the universal realm of illusion, but rather illuminates more clearly the paradoxical and infinitely precarious character of social existence.

Another concept of existentialist philosophy useful for our argument is what Martin Heidegger has called *das Man*. The German word is untranslatable literally into English. It is used in German in the same way that 'one' is used in English in such a sentence as 'One does not do that' ('*Man tut das nicht*'). The French word *on* conveys the same meaning, and José Ortega y Gasset has caught Heidegger's intention well in Spanish with his concept of *lo que se hace*. In other words, *Man* refers to a deliberately vague generality of human beings. It is not this man who will not do this, nor that man, nor you nor I – it is, in some way, all men, but so generally that it may just as well be nobody. It is in this vague sense that a child is told 'one does not pick one's nose in public'. The concrete child, with his concretely irritating nose, is subsumed under an anonymous generality that has no face – and yet bears down powerfully on the child's conduct. In fact (and this ought to give us a long pause), Heidegger's *Man* bears uncanny resemblance to what Mead has called the 'generalized other'.

In Heidegger's system of thought the concept of the *Man* is related to his discussion of authenticity and inauthenticity. To exist authentically is to live in full awareness of the unique, irreplaceable and incomparable quality of one's individuality. By contrast, inauthentic existence is to lose oneself in the anonymity of the *Man*, surrendering one's uniqueness to the socially constituted abstractions. This is especially important in the way one faces death. The truth of the matter is that it is always one single, solitary individual who dies. But society comforts the bereaved and those who are to die themselves by subsuming each death under general categories that appear to assuage its horror. A man dies, and we say 'Well, we all have to go someday'. This 'we all' is an exact rendition of the *Man* – it is everybody and thus nobody, and by putting ourselves under its generality we hide from ourselves the inevitable fact that we too shall die, singly and solitarily. Heidegger himself has referred to Tolstoy's story *The Death of Ivan Ilyitch* as the best literary expression of inauthenticity in the facing of death. As an illustration of authenticity to the point of torment we would submit Federico García Lorca's unforgettable poem about the death of a bullfighter, *Lament for Ignacio Sànchez Mejías*.

Heidegger's concept of *Man* is relevant for our view of society not so much in its normative as in its cognitive aspects. Under the aspect of 'bad faith' we have seen society as a mechanism to provide alibis from freedom. Under the aspect of the *Man* we see society as a defence against terror. Society provides us with taken-for-granted structures ʼwe could also speak here of the 'okay world') within which, as long as we follow the rules, we are shielded from the naked terrors of our condition. The 'okay world' provides routines and rituals through which these terrors are organized in such a way that we can face them with a measure of calm.

All rites of passage illustrate this function. The miracle of birth, the mystery of desire, the horror of death – all these are carefully camouflaged as we are led gently over one threshold after another, apparently in a natural and self-evident sequence; we all are born, lust and must die, and thus every one of us can be protected against the unthinkable wonder of these events. The *Man* enables us to live inauthentically by sealing up the metaphysical questions that our existence poses. We are surrounded by darkness on all sides as we rush through our brief span of life toward inevitable death. The agonized question 'Why?' that almost every man feels at some moment or other as he becomes conscious of his condition is quickly stifled by the cliché answers that society has available. Society provides us with religious systems and social rituals, ready made, that relieve us of such questioning. The 'world-taken-for-granted', the social world that tells us that everything is quite okay, is the location of our inauthenticity.

Let us take a man who wakes up at night from one of those nightmares in which one loses all sense of identity and location. Even in the moment of waking, the reality of one's own being and of one's world appears as a dream-like phantasmagorion that could vanish or be metamorphosed in the twinkling of an eye. One lies in bed in a sort of metaphysical paralysis, feeling oneself but one step removed from that annihilation that had loomed over one in the nightmare just passed. For a few moments of painfully clear consciousness one is at the point of almost smelling the slow approach of death and, with it, of nothingness. And then one gropes for a cigarette and, as the saying goes, 'comes back to reality'. One reminds oneself of one's name, address and occupation, of one's plans for the next day. One walks about one's house, full of proofs of past and present identity. One listens to the noises of the city. Perhaps one wakes up wife or children and is reassured by their annoyed protests. Soon one can laughingly dismiss the foolishness of what has just transpired, raid the refrigerator for a bite or the liquor closet for a nightcap, and go to sleep with the determination to dream of one's next promotion.

So far, so good. But what exactly is the 'reality' to which one has just returned? It is the 'reality' of one's socially constructed world, that 'okay world' in which metaphysical questions are always laughable unless they have been captured and castrated in taken-for-granted religious ritualism. The truth is that this 'reality' is a very precarious one indeed. Names, addresses, occupations and wives have a way of disappearing. All plans end in extinction. All houses eventually become empty. And even if we live all our lives without having to face the agonizing contingency of all we are and do, in the end we must return to that nightmare moment when we feel ourselves stripped of all names and all identities. What is more, we know this – which makes for the inauthenticity of our scurrying for shelter. Society gives us names to shield us from nothingness. It builds a world for us to live in and thus protects us from the chaos that surrounds us on all sides. It provides us with a language and with meanings that make this world believable. And it supplies a steady chorus of voices that confirm our belief and still our dormant doubts.

Again we would repeat in this slightly altered context what we have said before about 'bad faith'. It is correct that society, in its aspect of *Man*, is a conspiracy to bring about inauthentic existence. The walls of society are a Potemkin village erected in front of the abyss of being. They function to protect us from terror, to organize for us a cosmos of meaning within which our lives make sense.

But it is also true that authentic existence can take place only within society. All meanings are transmitted in social processes. One cannot be human, authentically *or* in-authentically, except in society. And the very avenues that lead to a wondering contemplation of being, be they religious or philosophical or aesthetic, have social locations. Just as society can be a flight from freedom or an occasion for it, society can bury our metaphysical quest or provide forms in which it can be pursued. We come up once more on the persistently Janus-faced paradox of our social existence. All the same, there can be but little doubt that society functions as alibi and as Potemkin village for more people than it functions for as an avenue of liberation. If we maintain that authenticity in society is possible, we are not thereby maintaining that most men are indeed making use of this possibility. Wherever we ourselves may be socially located, one look around us will tell us otherwise

With these observations we have come once more to the edge of ethical considerations that we want to postpone for another moment. We would stress at this point, however, that 'ecstasy', as we have defined it, has metaphysical as well as sociological significance. Only by stepping out of the taken-for-granted routines of society is it possible for us to confront the human condition without comforting mystifications. This does not mean that only the marginal man or the rebel can be authentic. It does mean that free-dom presupposes a certain liberation of consciousness. Whatever possibilities of freedom we may have, they cannot be realized if we continue to assume that the 'okay world' of society is the only world there is. Society provides us with warm, reasonably comfortable caves, in which we can huddle with our fellows, beating on the drums that drown out the howling hyenas of the surrounding darkness. 'Ecstasy' is the act of stepping outside the caves, alone, to face the night.

7

SOCIOLOGICAL MACHIAVELLIANISM
AND ETHICS

THE writer has elsewhere discussed at some length certain ethical implications of sociological thought. This was done with specific reference to a Christian view of man. The purpose of the present book, however, does not include the urging upon the reader of the writer's religious commitments. An invitation to profane subversiveness might suffice for one volume without adding to it a case for such corruption of the public verities as may come from sacred preoccupations. Nor can the discussion of ethical issues be more than very brief within the context of this book. However, we have touched upon ethical questions of some urgency in several places in the course of our argument, especially in the last chapter, and the reader has a right to demand at least an indication of how these questions could be answered.

Enough has been said in the preceding pages to justify the conclusion that sociological perspective is not conducive to an onward-and-upward outlook, but will rather lead to one degree or another of disenchantment with regard to the interpretations of social reality given in Sunday schools and civics classes. This is true no matter whether we proceed to the dramatic view of society discussed before or arrest ourselves at the more grimly deterministic models arrived at earlier. Viewing society as a carnival is, if anything, worse from the standpoint of the official ideologies than viewing it as a penitentiary. The Machiavellian possibilities of this sociological disenchantment are obvious. Although the positivistic dream of knowledge always leading to power is somewhat utopian, it remains true that clear-sightedness is conducive to the acquisition of control. Especially is this true with regard to clear-sightedness about social matters, as Machiavelli knew and taught.

Only he who understands the rules of the game is in a position to cheat. The secret of winning is insincerity. The man who plays all his roles sincerely, in the sense of un-reflected response to unscrutinized expectation, is incapable of 'ecstasy' – and, by the same token, quite safe from the viewpoint of those concerned with protecting the rules. We have tried to show how sociology can serve as a prelude to 'ecstasy', and, by implication, as a course in how to beat the system. Let no one quickly jump to the conclusion that such an ambition is always ethically reprehensible. That depends, after all, on how one evaluates the ethical status of the system in question. No one will object if the victims of a tyranny try to get away with a few tricks behind the tyrant's back. All the same, there is an ethically sinister possibility in knowing the machinery of rules. At least some of the popular mistrust of the social sciences is based on a correct if inarticulate hunch of this possibility. In this sense, every sociologist is a potential saboteur or swindler, as well as a putative helpmate of oppression.

As we pointed out much earlier in our argument, the social scientist shares this ethical predicament with his colleagues in the natural sciences, as the political use of nuclear physics has more than amply demonstrated in recent years. The prospect of politically controlled scientists working away on both sides of the Iron Curtain is not pleasant to behold. As the physicists are busy engineering the world's annihilation, the social scientists can be entrusted with the smaller mission of engineering the world's consent. Yet almost everyone will agree that these considerations cannot be concluded by placing the science of physics under an ethical anathema. The problem is not in the character of the science but in that of the scientist. The same thing holds for the sociologist and whatever powers he can muster, paltry though these may seem beside the demonic arsenal of the natural sciences.

Machiavellianism, be it political or sociological, is a way of looking, in itself ethically neutral. It becomes charged with negative ethical energy as it is applied by men without scruples or compassion. Friedrich Meinecke has shown convincingly in his history of political Machiavellianism that *raison d'état*, in the sense of the great Italian diagnostician of the body politic, is capable of being combined with the most serious ethical concern. Sociological Machiavellianism is no different. The life of Max Weber, for instance, is an object lesson in how ruthless sociological understanding can be combined with scrupulous search for the realization of ethical ideals. This does not change the sinister possibility of the tools of Machiavellian perception resting in the hands of men with inhuman purposes or with no purpose at all except to serve the powers that be. The applicability of sociological knowledge in the service of political propaganda and military planning in this country is chilling enough. It becomes nightmarish in the case of a totalitarian society. Nor is the spectacle of some of the uses of sociology in contemporary industrial management, public relations and advertising of great ethical edification. The fact that there are many sociologists who do not regard any of this as raising ethical questions is sufficient proof of the fact that sociological perspective does not *ipso facto* lead to a higher degree of ethical sensitivity. What is more, the totally cynical researcher is sometimes more accurate in his findings than a colleague burdened with scruples and a morally weak stomach, just because the latter may recoil from some of the things that he may find in the course of his research. One cannot even comfort oneself with the thought that it is the better social scientists (better in their scientific competence, that is) who are the more ethically concerned.

In this connexion, it is interesting to note how sociological understanding itself can become a vehicle of 'bad faith'. This occurs when such understanding becomes an alibi for responsibility. We have already alluded to this possibility in the first chapter in discussing the image of the sociologist as an impassive, uncommitted spectator. For example, a sociologist located in the South may start out with strong, personal values that repudiate the Southern racial system and he may seek to express these values by some form of social or political action. But then, after a while, he becomes an expert, *qua* sociologist, in racial matters. He now really feels that he understands the system. At this point, it may be observed in some cases, a different stance is adopted *vis-à-vis* the moral problems – that of the coolly scientific commentator. The sociologist now regards his act of understanding as constituting the sum total of his relationship to the phenomenon and as releasing him from any of those acts that would engage him personally. In such cases, the relationship between scientific objectivity and the subjectivity of the morally involved human being can be pictured in the analogy used by Sören Kierkegaard to describe Hegelian thought – one builds a magnificent palace, a wonder to behold, but goes on living in a hovel next to it. It is important to stress here that there is nothing ethically reprehensible in the role of scientific neutrality as such, and it is quite likely that in certain situations even the very much engaged sociologist may feel that this is the role in which he can make his best contribution. The ethical problem arises when this role is instituted *in lieu of* personal commitments in the total existence of the sociologist. In the latter case one is entitled to speak of 'bad faith', in Sartre's meaning of the term.

We concede to the critics of sociology that there are grounds here for genuine ethical concern. Nevertheless we contend that there are significant ethical possibilities directly grounded in sociological understanding. It must be made clear at once that we cannot accept or resurrect the old Comtian hope, still continuing in the Durkheimian tradition in French sociology, to the effect that sociological science will be able to come up with an objective morality (what the French would call a *science des moeurs*) on the basis of which a kind of secularist catechism could be established. Such hopes, some of which have found considerable resonance in America, are bound to be disappointed, because they fail to understand the fundamental disparateness of scientific and ethical judgements. Scientific methods are just as incapable of discovering what the good life is to be as they are incapable of coming upon freedom as an empirical phenomenon. To expect such feats of science is to misunderstand its peculiar genius. The disappointment that will follow makes it harder to see where the real human contributions of this genius can be found.

We instead maintain that sociology is capable of helping the individual acquiring it towards a certain humanization in his view of social reality. This is saying something with considerable care, since it has already been conceded that there is no necessity in this process. However, if one can accept the argument concerning sociological perspective in the preceding chapters, this humanization becomes at least an intellectual plausibility. Sociological understanding comes up again and again on the paradox of the ponderosity and the precariousness of society. To repeat, society defines man, and is in turn defined by man. This paradox touches essentially on the human condition as such. It would be very surprising indeed if this perspective had no ethical import at all, an assumption that could be made only if ethics is taken as a domain utterly divorced from the empirical world in which men live.

What we have called humanization here can be illustrated by three examples which, indeed, have a certain paradigmatic significance – the questions of race, homosexuality and capital punishment. In each of these it can be seen how sociological understanding can make a contribution on a superficial level of the objective clarification of issues. In actual fact, sociologists have made quite an important contribution on this level with regard to each of these questions. Sociologists have contributed greatly in debunking the mythologies associated with race, in uncovering the exploitative functions of these mythological beliefs, in showing more clearly how the racial system works in American society and thereby giving some ideas as to how the system could be effectively changed. In the case of homosexuality, sociologists have tended to leave the interpretation of the phenomenon itself to psychologists and psychiatrists, but they have accumulated data that show the distribution of the phenomenon and its social organization, thereby debunking the moralistic definition of homosexuality as the vice of a minute handful of degenerates and putting serious question marks behind the legal provisions concerning the phenomenon. In the case of capital punishment, sociologists have been able to demonstrate conclusively that the death penalty does not act as a deterrent to the commission of the crimes for which it is exacted and that the abolition of the death penalty does not lead to any of the frightful consequences predicted by its propagandists.

Without any question, these contributions have been of very great importance for an intelligent approach to public policy in these matters. They would be quite sufficient in themselves to justify the sociologist's claim that his activities have a moral value. However, we would argue that, in each of these three cases, sociology has a deeper contribution to make, one closely connected with what we have called humanization and directly rooted in the paradoxical understanding of social reality discussed previously.

Sociology shows man being that which society made him

and weakly, hesitatingly, sometimes passionately trying to be something else, something chosen by himself. Sociology uncovers the infinite precariousness of all socially assigned identities. Sociological perspective, as we understand it, is thus innately at odds with viewpoints that totally equate men with their socially assigned identities. To put this differently, the sociologist should be far too aware of the machinery of stage management to be taken in by the act being performed. He should know the acrobatics by which the actors have gotten into their costumes for any particular role, and this should make it very hard for him to give ontological status to the masquerade. The sociologist ought, therefore, to have difficulties with any set of categories that supply appellations to people – 'Negroes', 'whites', 'Caucasians', or, for that matter, 'Jews', 'Gentiles', 'Americans', 'Westerners'. In one way or another, with more or less malignancy, all such appellations become exercises in 'bad faith' as soon as they are charged with ontological implications. Sociology makes us understand that a 'Negro' is a person so designated by society, that this designation releases pressures that will tend to make him into the designated image, but also that these pressures are arbitrary, incomplete and, most importantly, reversible.

To deal with a human being exclusively *qua* 'Negro' is an act of 'bad faith', no matter whether these dealings are those of a racist or a racial liberal. In fact, it is worth stressing that liberals are often just as much caught in the fictions of socially taken-for-granted repertoires as are their political opponents; only they attach opposite valuations to these fictions. For that matter, those on the receiving end of negative identity assignments are very prone to accept the categories invented by their oppressors with the simple alteration of replacing the minus sign originally attached to the identity in question with a plus sign. Jewish reactions to anti-Semitism furnish classic illustrations of this process, with the Jewish counter-definitions of their own identity simply reversing the signs attached to the anti-Semitic categories without fundamentally challenging the categories themselves. To return to the Negro example, the process here takes the character of enjoining upon the Negro 'pride of race' in the place of his previous shame about it, thus building up a counter-formation of black racism that is but a shadow of its white prototype. Sociological understanding, by contrast, will make clear that the very concept of 'race' is nothing but a fiction to begin with, and perhaps help make clear that the real problem is how to be a human being. This is not to deny that counter-formations such as the ones mentioned may be functional in organizing resistance to oppression and may have a certain political validity much like other myths. All the same, they are rooted in 'bad faith', the corrosive power of which eventually exacts its toll as those who have painfully acquired 'pride of race' discover that their acquisition is a hollow one indeed.

Sociology is then conducive to an existential posture that is hard to reconcile with racial prejudice. This does not mean, unfortunately, that it excludes the latter. But the sociologist who holds on to such prejudice does so by virtue of a double dose of 'bad faith' – the 'bad faith' that is part and parcel of any racist stance, and his own special 'bad faith' by which he separates his sociological understanding from the rest of his existence in society. The sociologist who does not so segregate his intellect from his life, understanding the precarious manner in which social categories are concocted, will strive towards moral and political positions that are not hopelessly fixated on one set of categories taken with ultimate seriousness. In other words, he will take all socially assigned identities with a grain of salt, including his own.

The same logic applies to the case of homosexuality. The taken-for-granted contemporary Western attitude towards homosexuality, with its deposit in the *mores* and in the law, is based on the assumption that sexual roles are given by nature, that one set of sexual patterns is normal and healthy and desirable, another set abnormal, diseased and execrable. Again, sociological understanding will have to place a question mark behind this assumption. Sexual roles are constructed within the same general precariousness that marks the entire social fabric. Cross-cultural comparisons of sexual conduct bring home to us powerfully the near-infinite flexibility that men are capable of in organizing their lives in this area. What is normality and maturity in one culture, is pathology and regression in another. This relativization in the understanding of sexual roles does not, of course, free the individual from finding his own way morally. *That* would be another instance of 'bad faith', with the objective fact of relativity being taken as an alibi for the subjective necessity of finding those single decisive points at which one engages one's whole being. For instance, it is possible to be fully aware of the relativity and the precariousness of the ways by which men organize their sexuality, and yet commit oneself absolutely to one's own marriage. Such commitment, however, does not require any ontological underpinnings. It dares to choose and to act, refusing to push the burden of decision on nature or necessity.

The persecution of homosexuals fulfils the same function of 'bad faith' as racial prejudice or discrimination. In both cases, one's own shaky identity is guaranteed by the counter-image of the despised group. As Sartre has shown in his description of the anti-Semite, one legitimates oneself by hating the figure one has set up as the opposite of oneself. The white man despises the Negro and in that very act confirms his own identity as one entitled to show contempt. In the same way, one comes to believe one's own dubious virility as one spits upon the homosexual. If contemporary psychology has proved anything, it is the synthetic character of the virility of the *homme sexuel moyen*, that same erotic Babbit who likes to play the role of Torquemada in the persecution of sexual heresy. One does not have to develop great psychological sophistication to perceive the cold panic that lurks behind the gruff male demeanour of such types. The 'bad faith' in the act of persecution has the same roots that 'bad faith' has everywhere – the flight from one's own freedom, including that terrifying freedom (terrifying, at any rate, to the persecutor) of desiring a man instead of a woman. Again, it would be naïve to maintain that sociologists are not capable of such inauthenticity. However, we contend once more that sociological perspective on these phenomena will simultaneously relativize and humanize them. It will induce scepticism about the conceptual apparatus with which society assigns some human beings to darkness and others to light (including that modern modification of such an apparatus that identifies the darkness with 'pathology'). It will be conducive to the realization that all men struggle against powerful odds to define for themselves a constantly threatened and therefore all the more precious identity within that brief span of time that is their own.

Capital punishment can serve as the paradigm for the combination of 'bad faith' with inhumanity, for each step of this monstrous process, as it is still practised in America, is an act of 'bad faith', in which socially constructed roles are taken as alibis for personal cowardice and cruelty. The prosecuting attorney claims to suppress his sympathy to carry out his stern duty, as does the jury and the judge. Within the drama of a courtroom in which a capital case is tried, every one of those who prepare the eventual execution of the defendant is engaged in an act of deception – the deception that he is not acting as an individual, but only *qua* the role assigned to him in the edifice of legal fictions. The same pretence is carried further to the final act of the drama, the execution itself, in which those who order the killing, those who watch over it and those who physically perform it are all protected from personal accountability by the fiction that it is not really *they* who are engaged in these acts but anonymous beings representing 'the law', 'the state' or 'the will of the people'. So strong is the appeal of such fictions that people will even sympathize with the poor wardens or prison officials who 'must' do these cruel things in the line of duty. The excuse of such men that they 'have no choice' is the fundamental lie upon which all 'bad faith' rests. It is only quantitatively different from the same excuse proffered by the official murderers of the Nazi system of horrors. The judge who pleads necessity in sentencing a man to death is a liar, as is the warden who does so in executing the man, and the governor who refuses to stay the execution. The truth is that a judge can resign, that a warden can refuse to obey an order, and that a governor can stand up for humanity even against the law. The nightmare character of 'bad faith' in the case of capital punishment does not lie so much in the degree of deception (which can be matched elsewhere), but in the function carried by this deception – the killing of a human being with precise bestiality and in such a way that nobody need feel responsible.

The conviction of many people in our time that capital punishment is a monstrous inhumanity that lies beyond the limits of what is morally tolerable in a civilised community springs from a view of the human condition that one can certainly not equate with sociological perspective. It rests on a fundamental recognition of what is human and what is 'counter-human', to use a term employed by Martin Buber in the eloquent statement in which he deplored the execution of Adolf Eichmann. It is the same choice of being human that would, under certain circumstances and with ultimate reluctance, allow one to kill, but that would never allow one to torture. It is, in sum, the recognition of capital punishment as torture. This is not the place to argue how such an understanding of the human condition may come about. It certainly cannot be ascribed to sociology. However, we claim a more modest but nevertheless worthy task for the latter. Sociological understanding cannot by itself be a school of compassion, but it can illuminate the mystifications that commonly cover up pitilessness. The sociologist will understand that all social structures are conventions, shot through with fictions and with fraudulence. He will recognize that some of these conventions are useful and feel little inclination to change them. But he should have something to say when the conventions become instrumentalities of murder.

Perhaps enough has been said here to indicate the possibility that, if there is something like a sociological anthropology, there may also be something like a sociological humanism. Clearly sociology by itself cannot lead to humanism, as it cannot by itself produce an adequate anthropology (our own procedure in the last chapter should make the latter clear). But sociological understanding can be an important part of a certain sense of life that is peculiarly modern, that has its own genius of compassion and that can be the foundation of a genuine humanism. This humanism to which sociology can contribute is one that does not easily wave banners, that is suspicious of too much enthusiasm and too much certainty. It is an uneasy, uncertain, hesitant thing, aware of its own precariousness, circumspect in its moral assertions. But this does not mean that it cannot enter into passionate commitment at those points where its fundamental insights into human existence are touched upon. The three questions raised above would, indeed, serve well as primary indicators of where these points are. Before the tribunals that condemn some men to indignity because of their race or sexuality, or that condemn any man to death, this humanism becomes protest, resistance and rebellion. There are, of course, other points at which compassion can become the starting point of revolution against systems of inhumanity sustained by myth. However, on most other issues, in which human dignity is less crucially involved, the sociological humanism that we are suggesting is likely to adopt a more ironic stance. And a few final remarks on that may be in order here.

Sociological understanding leads to a considerable measure of disenchantment. The disenchanted man is a poor risk for both conservative and revolutionary movements; for the former because he does not possess the requisite amount of credulity in the ideologies of the *status quo*, for the latter because he will be sceptical about the utopian myths that invariably form the nurture of revolutionaries. Such unemployability in the cadres of either present or future régimes need not, however, leave the disenchanted man in the posture of alienated cynicism. It may do that, to be sure. And we find just such postures among some younger sociologists in America, who find themselves driven to radical diagnosis of society without finding in themselves the capacity for radical political commitments. This leaves them with no place to go except to a sort of masochistic cult of debunkers who reassure each other that things could not possibly be worse. We contend that this cynical stance is in itself naïve and often enough grounded more in a lack of historical perspective than anything else. Cynicism about society is not the only option besides a credulous conformity to this social aeon or a credulous looking-forward to the one that is to come.

Another option is what we regard as the most plausible one to result from sociological understanding, one that can combine compassion, limited commitment and a sense of the comic in man's social carnival. This will lead to a posture *vis-à-vis* society based on a perception of the latter as essentially a comedy, in which men parade up and down with their gaudy costumes, change hats and titles, hit each other with the sticks they have or the ones they can persuade their fellow actors to believe in. Such a comic perspective does not overlook the fact that non-existent sticks can draw real blood, but it will not from this fact fall into the fallacy of mistaking the Potemkin village for the City of God. If one views society as a comedy, one will not hesitate to cheat, especially if by cheating one can alleviate a little pain here or make life a little brighter there. One will refuse to take seriously the rules of the game, except insofar as these rules protect real human beings and foster real human values. Sociological Machiavellianism is thus the very opposite of cynical opportunism. It is the way in which freedom can realize itself in social action.

8

SOCIOLOGY AS A
HUMANISTIC DISCIPLINE

SOCIOLOGY, from the beginning, understood itself as a science. Very early in our argument we discussed some methodological consequences of this self-understanding. In these final remarks, we are not concerned with methodology but rather with the human implications of having an academic discipline such as sociology. We have tried to depict in previous chapters the way in which sociological perspective helps to illuminate man's social existence. In the last digression we briefly asked what the ethical implications of such perspective may be. We now conclude by looking once more at sociology as one discipline among many in that particular corner of the social carnival that we call scholarship.

One very important thing that many sociologists can learn from their fellow scientists in the natural sciences is a certain sense of play with regard to their discipline. Natural scientists, on the whole, have with age acquired a degree of sophistication about their methods that allows them to see the latter as relative and limited in scope. Social scientists still tend to take their discipline with grim humourlessness, invoking terms such as 'empirical', 'data', 'validity' or even 'facts' as a voodoo magician might his most cherished hobgoblins. As the social sciences move from their enthusiastic puberty to a mellower maturity, a similar degree of detachment from one's own game may be expected and, indeed, can already be found. One can then understand sociology as one game among many, significant but hardly the last word about human life, and one can afford not only tolerance but even an interest in other people's epistemological entertainments.

Such a mellowing in self-understanding is in itself of human significance. It might even be said that the mere presence in an intellectual discipline of ironical scepticism concerning its own undertakings is a mark of its humanistic character. This is all the more important for the social sciences, dealing as they do with the peculiarly ludicrous phenomena that constitute the 'human comedy' of society. Indeed, an argument could be made that the social scientist who does not perceive this comic dimension of social reality is going to miss essential features of it. One cannot fully grasp the political world unless one understands it as a confidence game, or the stratification system unless one sees its character as a costume party. One cannot achieve a sociological perception of religious institutions unless one recalls how as a child one put on masks and frightened the wits out of one's contemporaries by the simple expedient of saying 'boo'. No one can understand any aspect of the erotic who does not grasp its fundamental quality as being that of an *opéra bouffe* (a point one should especially emphasize to serious young sociologists teaching courses on 'courtship, marriage and the family' with an unsmiling seriousness that hardly fits the study of a field every aspect of which hangs, so to speak, from that part of the human anatomy that is most difficult to take seriously). And a sociologist cannot understand the law who does not recollect the jurisprudence of a certain Queen in *Alice in Wonderland*. These remarks, needless to say, are not meant to denigrate the serious study of society, but simply to suggest that such study itself will profit greatly from those insights that one can obtain only while laughing.

Sociology will be especially well advised not to fixate itself in an attitude of humourless scientism that is blind and deaf to the buffoonery of the social spectacle. If sociology does that, it may find that it has acquired a foolproof methodology, only to lose the world of phenomena that it originally set out to explore – a fate as sad as that of the magician who has finally found the formula that will release the mighty *jinn* from the bottle, but cannot recollect what it was that he wanted to ask of the *jinn* in the first place. However, while eschewing scientism, the sociologist will be able to discover human values that are endemic to scientific procedures in both the social and the natural sciences. Such values are humility before the immense richness of the world one is investigating, an effacement of self in the search for understanding, honesty and precision in method, respect for findings honestly arrived at, patience and a willingness to be proven wrong and to revise one's theories, and, last but not least, the community of other individuals sharing these values.

The scientific procedures used by the sociologist imply some specific values that are peculiar to this discipline. One such value is the careful attention to matters that other scholars might consider pedestrian and unworthy of the dignity of being objects of scientific investigation – something one might almost call a democratic focus of interest in the sociological approach. Everything that human beings are or do, no matter how common-place, can become significant for sociological research. Another such peculiar value is inherent in the sociologist's necessity to listen to others without volunteering his own views. The art of listening, quietly and with full attention, is something that any sociologist must acquire if he is to engage in empirical studies. While one should not exaggerate the importance of what is often nothing more than a research technique, there is a human significance at least potentially present in such conduct, especially in our nervous and garrulous age in which almost nobody finds the time to listen with concentration. Finally, there is a peculiar human value in the sociologist's responsibility for evaluating his findings, as far as he is psychologically able, without regard to his own prejudices, likes or dislikes, hopes or fears. This responsibility, of course, the sociologist shares with other scientists. But it is especially difficult to exercise in a discipline that touches so closely on the human passions. It is evident that this goal is not always achieved, but in the very effort lies a moral significance not to be taken lightly. This becomes particularly appealing when one compares the sociologist's concern for listening to the world, without immediately shouting back his own formulations of what is good and what is bad, with the procedures of normative disciplines, such as theology or jurisprudence, where one meets with the constant compulsion to squeeze reality into the narrow frame of one's value judgements. Sociology appears by comparison as standing in an apostolic succession from the Cartesian quest for 'clear and distinct perception'.

In addition to these human values that are inherent in the scientific enterprise of sociology itself, the discipline has other traits that assign it to the immediate vicinity of the humanities if they do not, indeed, indicate that it belongs fully with them. In the preceding chapters we have been at pains to explicate these traits, all of which could be summarized by saying that sociology is vitally concerned with what is, after all, the principal subject matter of the humanities – the human condition itself. Just because the social is such a crucial dimension of man's existence, sociology comes time and again on the fundamental question of what it means to be a man and what it means to be a man in a particular situation. This question may often be obscured by the paraphernalia of scientific research and by the bloodless vocabulary that sociology has developed in its desire to legitimate its own scientific status. But sociology's data are cut so close from the living marrow of human life that this question comes through again and again, at least for those sociologists who are sensitive to the human significance of what they are doing. Such sensitivity, as we have argued, is not just an adiaphoron that a sociologist may possess in addition to his properly professional qualifications (such as a good ear in music or a knowing palate for food), but has direct bearing upon sociological perception itself.

Such an understanding of the humanistic place of sociology implies an openness of mind and a catholicity of vision. It should be readily conceded that such a posture may be acquired at the cost of rigorously closed logic in the task of sociological system-building. Our own argument can serve as an embarrassing illustration of this weakness. The reasoning pursued in Chapters 4 and 5 of this book could be logically fixated in a theoretical system of sociologism (that is, a system that interprets all of human reality consistently and exclusively in sociological terms, recognizing no other causal factors within its preserve and allowing for no loopholes whatever in its causal construction). Such a system is neat, even aesthetically pleasing. Its logic is one-dimensional and closed within itself. That this sort of intellectual edifice is inviting to many orderly minds is demonstrated by the appeal that positivism in all its forms has had since its inception. The appeal of Marxism and Freudianism has very similar roots. To conduct a sociological argument and then to veer off from its seemingly compelling sociologistic conclusion must give the appearance of being inconsequential and less than rigorous in one's thinking, as the reader may have felt as our argument began to backtrack in Chapter 6. All this can readily be admitted – and yet followed by the contention that the inconsequence is due not to the perversity of the observer's reasoning but to the paradoxical many-sidedness of life itself, that same life he is committed to observe. Such openness to the immense richness of human life makes the leaden consequences of sociologism impossible to sustain and forces the sociologist to permit 'holes' in the closed walls of his theoretical scheme, openings through which other possible horizons can be perceived.

Openness to the humanistic scope of sociology further implies an ongoing communication with other disciplines that are vitally concerned with exploring the human condition. The most important of these are history and philosophy. The foolishness of some sociological work, especially in America, could be easily avoided by a measure of literacy in these two areas. While most sociologists, by temperament perhaps or by professional specialization, will be concerned mainly with contemporary events, disregard of the historical dimension is an offence not only against the classic Western ideal of the civilized man but against sociological reasoning itself – namely, that part of it that deals with the central phenomenon of predefinition. A humanistic understanding of sociology leads to an almost symbiotic relationship with history, if not to a self-conception of sociology as being itself a historical discipline (a notion still alien to most American sociologists but quite common in Europe). As to philosophical literacy, it would not only prevent the methodological naïveté of some sociologists, but would also be conducive to a more adequate grasp of the phenomena themselves that the sociologist wishes to investigate. None of this should be construed as a denigration of statistical techniques and other equipment burrowed by sociology from quite definitely non-humanistic sources. But the use of these will be more sophisticated and also (if one may say this) more civilized if it occurs against a background of humanistic awareness.

The notion of humanism has been closely connected with that of intellectual liberation since the Renaissance. Enough has been said in preceding pages to serve as a substantiation of the claim that sociology stands rightfully in this tradition. In conclusion, however, we may ask in what way the sociological enterprise in America (itself now constituting a social institution and a professional sub-culture) can lend itself to this humanistic mission. This question is not new and has been asked incisively by sociologists such as Florian Znaniecki, Robert Lynd, Edward Shils and others. But it is important enough not to omit before this argument is brought to a close.

An alchemist locked up by a predatory prince who needs gold and needs it quickly will have had little chance to interest his employer in the lofty symbolism of the Philosopher's Stone. Sociologists employed in many agencies of government and branches of industry will often be in roughly the same position. It is not easy to introduce a humanistic dimension into research designed to determine the optimum crew composition of a bomber aircraft, or to discover the factors that will induce somnambulant housewives in a supermarket to reach for one brand of baking powder as against another, or to advise personnel managers about the best procedures to undermine union influence in a factory. Although sociologists employed in such useful activities may prove to their own satisfaction that there is nothing ethically questionable about these applications of their craft, to look upon them as humane endeavours would require something of a *tour de force* in ideologizing. On the other hand, one should not too summarily dismiss the possibility that a certain humane emphasis can nevertheless result from the application of the social sciences to governmental or industrial operations. For example, the place of sociologists in various programmes of public health, welfare planning, urban redevelopment or in governmental agencies concerned with the eradication of racial discrimination should prevent us from concluding too rapidly that governmental employment *must* mean for the sociologist a soulless captivity to political pragmatism. Even in industry a case might be made that the most intelligent and forward-looking thinking in management (especially in the area of personnel management) has profited greatly from sociological contributions.

If the sociologist can be considered a Machiavellian figure, then his talents can be employed in both humanly nefarious and humanly liberating enterprises. If a somewhat colourful metaphor may be allowed here, one can think of the sociologist as a *condottiere* of social perception. Some *condottieri* fight for the oppressors of men, others for their liberators. Especially if one looks around beyond the frontiers of America as well as within them, one can find enough grounds to believe that there is a place in today's world for the latter type of *condottiere*. And the very detachment of sociological Machiavellianism is a not inconsiderable contribution in situations where men are torn by conflicting fanaticisms that have one important thing in common – their ideological befuddlement about the nature of society. To be motivated by human needs rather than by grandiose political programmes, to commit oneself selectively and economically rather than to consecrate oneself to a totalitarian faith, to be compassionate and sceptical at the same time, to seek to understand without bias – all these are existential possibilities of the sociological enterprise that can hardly be over-rated in many situations in the contemporary world. In this way, sociology can attain to the dignity of political relevance as well, not because it has a particular political ideology of its own to offer, but just because it has not. Especially to those who have become disillusioned with the more fervent political eschatologies of our era, sociology can be of assistance in pointing to possibilities of political engagement that do not demand the sacrifice of one's soul and of one's sense of humour.

It remains true in America, however, that most sociologists continue to be employed in academic institutions. It is likely that this will continue to be the case in the foreseeable future. Any reflections about the humanistic potential of sociology must, therefore, face up to the academic context in which most of American sociology is located. The notion of some academicians that only those who obtain their salaries from political and economic organizations get *les mains sales* is a preposterous one, in itself an ideology that serves to legitimate the academician's own position. For one thing, the economics of scientific research today is of such a nature that the academic world itself is permeated with the pragmatic interests of these extraneous organizations. Even though there are many sociologists who do not ride themselves on the gravy train of governmental or business opulence (mostly to their intense chagrin), the technique known to academic administrators as the 'freeing of funds' (more nastily referred to as the 'cigar-box method') ensures that the more esoteric professorial pursuits can also be nourished from the crumbs that fall off said gravy train.

However, even if one concentrates on the academic process proper, there is little justification for nose-thumbing on the part of the academically employed sociologist. The rat race of the university is frequently even more savage than the proverbial one on Madison Avenue, if only because its viciousness is camouflaged by scholarly courtesies and dedication to pedagogic idealism. When one has tried for a decade to get out of a third-string junior college to one of the prestige universities, or when one has tried in one of the latter to make an associate professorship for the same length of time, the humanistic impulse of sociology will have undergone at least as much strain as it would under the aegis of non-academic employers. One will write those things that have a chance to be published in the right places, one will try to meet those people who dwell close to the mainsprings of academic patronage, one will fill the gaps in one's vita with the same political assiduousness as any junior executive on the make, and one will quietly detest one's colleagues and one's students with the intensity of shared imprisonment. So much for academic pretentiousness.

The fact remains that if sociology has a humanistic character, that character will have to manifest itself within the academic milieu, if only for statistical reasons. We would contend that, despite the uncomplimentary remarks just made, this is a realistic possibility. The university is much like the church in its susceptibility to seduction by the powerful of this world. But university people, like churchmen, develop a guilt complex after the seduction has been accomplished. The old Western tradition of the university as a place of freedom and of truth, a tradition fought for with blood as well as ink, has a way of reasserting its claims before an uneasy conscience. It is within this persistent academic tradition that the humanistic impulse in sociology can find its living space in our contemporary situation.

It is obvious that a difference exists in the problems faced in this regard in a graduate school concerned with the training of a new generation of sociologists and in an undergraduate situation. In the former case, the problem is relatively easy. Naturally the writer would feel that the conception of sociology developed here should find its place in the 'formation' of future sociologists. The implications of what has been said about the humanistic dimension of sociology for graduate curricula in the discipline are obvious. This is not the place to develop them. Suffice it to say that humanistic literacy waxing at the expense of technological professionalism is the course that we envisage in this connexion. Evidently one's conception of sociology as a discipline will decide one's views as to how sociologists should be educated. But whatever this conception may be, it will be relevant to only a limited number of students. Not everyone, blessedly, can become a full-fledged sociologist. The one who does, if our argument is accepted, will have to pay the price of disenchantment and find his way in a world that lives on myth. We have said enough to indicate how we believe this to be possible.

The problem is obviously a different one in an undergraduate college. If a sociologist teaches in such a situation (most sociologists do), very few of his students will go on to graduate schools to study his particular field. It is even probable that very few of the sociology majors will, going on instead to social work, journalism, business administration, or any number of other occupations in which a 'sociological background' has been deemed useful. A sociologist teaching in many an average college, looking over his classes of young men and women desperately intent on social mobility, seeing them fight their way upward through the credit system and argue over grades with pertinacity, understanding that they could not care less if he read the phone directory to them in class as long as three credit hours could be added to the ledger at the end of the semester – such a sociologist will have to wonder sooner or later what sort of vocation it is that he is exercising. Even a sociologist teaching in a more genteel setting, providing intellectual pastime to those whose status is a foregone conclusion and whose education is the privilege rather than the instrumentality of such status, may well come to question what point there is to sociology, of all fields, in this situation. Of course, in state universities as well as in Ivy League colleges there are always the few students who really care, really understand, and one can always teach with only those in mind. This, however, is frustrating in the long run, especially if one has some doubts about the pedagogic usefulness of what one is teaching. And that is precisely the question that a morally sensitive sociologist ought to ask himself in an undergraduate situation.

196

The problem of teaching students who come to college because they need a degree to be hired by the corporation of their choice or because this is what is expected of them in a certain social position is shared by the sociologist with all his colleagues in other fields. We cannot pursue it here. There is, however, a peculiar problem for the sociologist that is directly related to the debunking, disenchanting character of sociology that we discussed before. It may well be asked with what right he peddles such dangerous intellectual merchandise among young minds that, more likely than not, will misunderstand and misapply the perspective he seeks to communicate. It is one thing to dispense the sociological poison to such graduate students as have already committed themselves to full-time addiction and who, in the course of intensive study, can be led to understand the therapeutic possibilities present in that poison. It is another thing to sprinkle it liberally among those who have no chance or inclination to proceed to that point of deeper understanding. What right does any man have to shake the taken-for-granted beliefs of others? Why educate young people to see the precariousness of things they had assumed to be absolutely solid? Why introduce them to the subtle erosion of critical thought? Why, in sum, not leave them alone?

Evidently at least part of the answer lies in the responsibility and the skill of the teacher. One will not address a freshman class as one would a graduate seminar. Another partial answer could be given by saying that the taken-for-granted structures are far too solidly entrenched in consciousness to be that easily shaken by, say, a couple of sophomore courses. 'Culture shock' is not induced that readily. Most people who are not prepared for this sort of relativization of their taken-for-granted world view will not allow themselves to face its implications fully and will instead look upon it as an interesting intellectual game to be played in their sociology class, very much as one may play the game of discussing whether an object is there when one is not looking at it in a philosophy class – that is, one will play the game without for a moment seriously doubting the ultimate validity of one's previous common-sense perspective. This partial answer has its merits too, but it will hardly do as a justification of the sociologist's teaching, if only because it applies only to the degree that this teaching fails to achieve its purpose.

We maintain that the teaching of sociology is justified insofar as a liberal education is assumed to have a more than etymological connexion with intellectual liberation. Where this assumption does not exist, where education is understood in purely technical or professional terms, let sociology be eliminated from the curriculum. It will only interfere with the smooth operation of the latter, provided, of course, that sociology has not also been emasculated in accordance with the educational ethos prevailing in such situations. Where, however, the assumption still holds, sociology is justified by the belief that it is better to be conscious than unconscious and that consciousness is a condition of freedom. To attain a greater measure of awareness, and with it of freedom, entails a certain amount of suffering and even risk. An educational process that would avoid this becomes simple technical training and ceases to have any relationship to the civilizing of the mind. We contend that it is part of a civilized mind in our age to have come in touch with the peculiarly modern, peculiarly timely form of critical thought that we call sociology. Even those who do not find in this intellectual pursuit their own particular demon, as Weber put it, will by this contact have become a little less stolid in their prejudices, a little more careful in their own commitments and a little more sceptical about the commitments of others – and perhaps a little more compassionate in their journeys through society.

Let us return once more to the image of the puppet theatre that our argument conjured up before. We see the puppets dancing on their miniature stage, moving up and down as the strings pull them around, following the prescribed course of their various little parts. We learn to understand the logic of this theatre and we find ourselves in its motions. We locate ourselves in society and thus recognize our own position as we hang from its subtle strings. For a moment we see ourselves as puppets indeed. But then we grasp a decisive difference between the puppet theatre and our own drama. Unlike the puppets, we have the possibility of stopping in our movements, looking up and perceiving the machinery by which we have been moved. In this act lies the first step towards freedom. And in this same act we find the conclusive justification of sociology as a humanistic discipline.

BIBLIOGRAPHICAL COMMENTS

THIS book has been an invitation to attend a certain party. In issuing such an invitation, it is not customary to include a full dossier on all the people that the putative guest will find at the party. All the same, the latter will want to know a little more about these people, or at least about where he can go to find out. It would be preposterous to conclude this book with an enormous bibliography on the various branches of sociology touched upon in the text. However, some bibliographical information should be given to the reader who has been intrigued by the invitation at least to the extent of wanting to look into the matter a little further. The purpose of these bibliographical comments is simply to suggest some points at which such further inquiry could profitably begin. Also, various names have been dropped in the text without much explanation. These comments will pick up these names and introduce them to the reader a little more fully. Obviously it is up to the reader how far he wants to go in response to this invitation. He has already been warned that such an undertaking is not without risk.

The interested reader should have little difficulty in obtaining the majority of the books referred to in the following bibliographical comments. Whenever possible recent English publications have been cited; these, and most of the American works referred to, are obtainable at academic and general booksellers. A few of the older works are out of print, but can generally be obtained at large public libraries.

Unless the reader is a student, and perhaps even then, he may have a deeply rooted aversion to the use of a textbook in approaching a new subject. Often enough this aversion is fully justified. There are notable exceptions, though. A classic textbook in sociology is Robert M. McIver, *Society* (New York, Farrar and Rinehart, 1937), still well worth reading. Among recent efforts in textbook composition particularly lucid ones are Ely Chinoy, *Society* (New York, Random House, 1961) and Harry M. Johnson, *Sociology – A Systematic Introduction* (London, Routledge & Kegan Paul, 1961).

Max Weber (1863-1920) was one of the giants in the development of sociology, very much rooted in the intellectual milieu of Germany in his time but with a continuing influence far beyond the frontiers of his native country. Weber's approach to sociology is especially characterized by its philosophical sophistication, its solid foundation in historical scholarship and its staggering scope in terms of the numbers of cultures drawn into the various carefully argued analyses of Weber's opus. For an eloquent statement of Weber's conception of sociology as a scientific discipline the reader may turn to the essay 'Science as a Vocation', found in an English translation in Edward A. Shils and Henry A. Finch (trs. and eds.), *The Methodology of the Social Sciences* (Chicago, Free Press, 1949). In the same volume will be found some other important formulations of Weber's understanding of scientific method. Max Weber, *Basic Concepts in Sociology*, translated by H. P. Secher (London, Peter Owen, 1962) is a small but useful volume. An extremely valuable work dealing with Weber's empirical studies is Reinhard Bendix, *Max Weber: An Intellectual Portrait* (London, Heinemann, 1960).

Alfred Schuetz (1899-1959) was a philosopher of the

phenomenological school who devoted most of his lifework to the philosophical foundations of sociology as a science. Born in Austria, he left that country after its occupation by the Nazis and taught until his death at the New School for Social Research in New York. His influence on contemporary sociologists is as yet limited but will undoubtedly grow as his work becomes more accessible. A Dutch publisher, Martinus Nijhoff of The Hague, is now preparing a three-volume English edition of Schuetz's opus.

CONCERNING CHAPTER 2

A concise statement of Albert Salomon's understanding of the historical origins of sociology in France can be found in his book *The Tyranny of Progress* (New York, Noonday Press, 1955). Paul Radin was an American anthropologist with a long list of works on primitive society. A good point at which to begin looking into this opus would be his *Primitive Man as Philosopher* (New York, D. Appleton and Co., 1927). A standard work on the history of social thought, including sociology, is Howard Becker and Harry E. Barnes, *Social Thought from Lore to Science*, available in a three-volume paperback edition (New York, Dover Publications, 1961). A briefer introduction to the development of sociological thought proper is Nicholas S. Timasheff, *Sociological Theory* (Garden City, N.Y., Doubleday and Co., 1955).

Of studies mentioned in connexion with the sociological penchant for looking under rugs, the reader may want to look into Robert Presthus, *Men at the Top: A Study in Community Power* (New York, Oxford University Press, 1964), and at two English studies, Margaret Stacey, *Tradition and Change: A Study of Banbury* (London, Oxford University Press, 1960), and Norbert Elias and John L. Scotson, *The Established and the Outsiders* (London,

Frank Cass and Co., 1965). Paul M. Harrison, *Authority and Power in the Free Church Tradition* (Princeton, Princeton University Press, 1959), deals with the relationship between Protestant denominational organization and bureaucracy.

Weber's book *The Protestant Ethic and the Spirit of Capitalism* is one of the most important works ever published in sociology. It has not only had a decisive influence on the development of sociology itself, but has been very influential among historians concerned with the relationships of economic and cultural history in the modern West. On the latter question, Weber's thesis on Protestantism and capitalism has been an important element of the critique of Marxist economic determinism. First published in German in 1905, the work was published in English in 1930 by George Allen and Unwin, London, and Charles Scribner's Sons, New York. It is also available in a paperback edition (London, George Allen and Unwin, 1965). E. Fischoff, *Max Weber: The Sociology of Religion* (London, Methuen and Co., 1965), contains much important translated material, and has an interesting introduction by Talcott Parsons.

Émile Durkheim (1858-1917) was the most important French sociologist in the formative period of the discipline. Around the journal *Année sociologique* he gathered a large school of disciples, working in various sectors of the social sciences, a school that continued after his death. Durkheim's sociology stands in the tradition of Comtian positivism, is characterized by its emphasis on the non-subjective quality of social phenomena, its pioneering use of statistical data, its close alliance with ethnological work and its ideological affinity with the ethos of French republicanism. A lucid view of Durkheim's understanding of sociology can be obtained from his programmatic work *The Rules of Sociological Method* (Chicago, Free Press, 1950). A good overall treatment of Durkheim's work can be found in Harry Alpert, *Émile Durkheim and his Sociology* (New York, Russell and Russell Inc., 1961).

Robert K. Merton of Columbia University is, along with Talcott Parsons of Harvard, the outstanding theorist in contemporary American sociology. Merton's discussion of 'manifest' and 'latent' functions, along with other important statements of what he considers to be the functionalist approach to society, will be found in his *Social Theory and Social Structure* (Chicago, The Free Press of Glencoe, 1957).

The concept of ideology was coined by the French philosopher Destutt de Tracy and used in a more strictly sociological sense by Marx. In subsequent sociology, however, the concept was greatly modified from its Marxian prototype. Vilfredo Pareto (1848-1923), an Italian scholar who spent many years of his life teaching in Switzerland, is notable for the construction of a sociological system essentially based on a concept of ideology. Pareto's main opus is available in a four-volume English edition under the title *The Mind and Society* (New York, Harcourt, Brace and Co., 1935), a formidable chunk of cerebration to chew through but well worth the effort of a courageous reader. Pareto's work was introduced to American sociologists by Talcott Parsons in the latter's *The Structure of Social Action* (Chicago, Free Press, 1949). The reader who does not feel up to Pareto's luxurious prose itself will find a concise discussion of Pareto's most important ideas in Parsons's volume. The most important use of the concept of ideology in contemporary sociology is in the so-called sociology of knowledge, of which more is said in Chapter 5 of this book. The basic reference for this is Karl Mannheim's book *Ideology and Utopia*, published in a paperback edition by Routledge & Kegan Paul, London, 1960.

Thorstein Veblen (1857-1929) was one of the most colourful of the early figures in American sociology. His approach to sociology is characterized by its merciless debunking orientation, its concentration on economic factors in social development and a strong affinity with radical critiques of capitalist society. *The Theory of the Leisure Class* (paperback edition, New York, Mentor Books, 1953), originally an analysis of the American upper classes, has become Veblen's most influential work in general sociological theory. *The Higher Learning in America* (New York, B. W. Huebsch, 1918) is one of the most bitter sociological treatises ever written, literally dripping venom on every page, an eloquent testimony to Veblen's savage disillusionment with American university life.

The so-called 'Chicago School' was a movement of sociologists, grouped around Robert Park at the University of Chicago, that produced a large number of studies of urban society in the 1920s. It has had a lasting influence on urban sociology, community studies and sociological analysis of occupations. A good discussion of Park's approach to sociology may be found in Maurice R. Stein, *The Eclipse of Community* (paperback edition, New York, Harper and Row, 1964). A good general discussion of the 'Chicago School' may be found in John Madge, *The Origins of Scientific Sociology* (London, Tavistock Publications, 1963). The most famous American community studies are the two investigations into the life and ways of Muncie, Indiana, undertaken by Robert S. and Helen Lynd just before and just after the Great Depression – *Middletown* (New York, Harcourt, Brace and Co., 1929) and *Middletown in Transition* (New York, Harcourt, Brace and Co., 1937). A worthy successor of the Lynds' determined digging below the veneer of community ideology is Arthur J. Vidich and Joseph Bensman, *Small Town in Mass Society*

(Princeton, Princeton University Press, 1958; paperback edition by Doubleday Anchor Books, 1960), a wry journey through the underside of the social structure of a rural community in upstate New York.

Daniel Lerner teaches sociology at the Massachusetts Institute of Technology. His book with Lucille W. Pevsner, *The Passing of Traditional Society* (paperback edition, Chicago, Free Press, 1964), not only gives an excellent sociological view of current developments in the Middle East, but is of more general importance in its theory of the modern mind as emerging out of older traditional patterns.

CONCERNING CHAPTER 3

For better or for worse, the writer of this book must bear the blame for most of the ideas in this digression. However, his thinking on these matters has been very strongly influenced by the teaching of Alfred Schuetz and some ideas of Maurice Halbwachs.

CONCERNING CHAPTER 4

There is a huge literature on stratification in contemporary sociology. The reader might profitably begin looking into this literature by turning to an anthology – Reinhard Bendix and Seymour M. Lipset (eds.), *Class, Status and Power* (London, Routledge & Kegan Paul, 1954). The same authors' study, *Social Mobility in Industrial Society* (London, Heinemann, 1959) can be read in conjunction with D. V. Glass, *Social Mobility in Britain* (London, Routledge & Kegan Paul, 1954). Ralf Dahrendorf, *Class and Class Conflict in Industrial Society* (London, Routledge & Kegan Paul, 1959) is valuable both for the author's own theory of class conflict and for his sociological critique of Marx.

William I. Thomas was an American sociologist who, together with Florian Znaniecki, wrote the gigantic study of immigration entitled *The Polish Peasant in Europe and America*, the first part of which was published in 1919 (Boston, Richard G. Badger). Many of Thomas's contributions to sociological theory are found in footnotes and appendices of this monumental opus, a charming if not always convenient place for such contributions. This study, incidentally, marked the beginning of a primarily empirical period in American sociology (the narrowness of which cannot be blamed on either Thomas or Znaniecki). An interesting but polemical criticism of this extreme empiricism is contained in a chapter in C. W. Mills, *The Sociological Imagination* (New York, Oxford University Press, 1959).

Arnold Gehlen is a contemporary German social scientist and philosopher. Together with Helmut Schelsky he has been influential in the renascence of sociology in Germany following World War II. As far as this writer knows, none of his works is available in English at this time.

CONCERNING CHAPTER 5

Charles Horton Cooley was one of the early American sociologists, theoretical in orientation and strongly influenced by European thought. His most important work is *Human Nature and the Social Order* (New York, Charles Scribner's Sons, 1922). George Herbert Mead was probably the most important figure in the development of American social psychology. He taught for many years at the University of Chicago and his most important work was published shortly after his death – *Mind, Self and Society* (Chicago, University of Chicago Press, 1934). Mead is a formidable author to grapple with, but essential for an understanding of the foundations of role theory. For recent statements of role theory and its more general implications the reader may turn to the following works: Hans H. Gerth and C. Wright Mills, *Character and Social Structure* (New York, Harcourt, Brace and Co., 1953); Erving Goffman, *The Presentation of Self in Everyday Life* (Garden City, N.Y., Doubleday Anchor, 1959); Anselm L. Strauss, *Mirrors and Masks* (New York, The Free Press of Glencoe, 1959). Goffman's highly suggestive analysis of the pressures of group therapy will be found in his recent book *Asylums* (Garden City, N.Y., Doubleday Anchor, 1961).

Max Scheler was a German philosopher, also greatly influenced by phenomenology, who developed the idea of a sociology of knowledge (which he called *Wissenssoziologie*) in the 1920s. Although some of Scheler's works are now available in English translation, this is not the case with those most important for this subject. Karl Mannheim was a sociologist greatly influenced by Scheler. His most important work, *Ideology and Utopia*, already mentioned above, was published in Germany. Mannheim left Germany for England after the advent of Nazism and became very influential there in the academic establishment of sociology. The reader will find an excellent introduction to the sociology of knowledge in the work of Robert Merton cited before. A more discursive treatment may be found in Werner Stark, *The Sociology of Knowledge* (London, Routledge & Kegan Paul, 1958).

Helmut Schelsky, who now teaches at the University of Munster, has written a number of articles on the religious consciousness of modern man that not only roused the interest of social scientists but created a furore in theological circles in Germany. Unfortunately these are not available in English. Thomas Luckmann teaches at the University of Frankfurt, and a very suggestive German work of his on the same subject is to be published shortly in English.

Talcott Parsons of Harvard University has become the founder of the most prominent school of sociological theory in America today. Parsons has set out to integrate the classic theories of European sociology with the theoretical approaches of other social-scientific disciplines, notably anthropology, psychology and economics. Parsons's system of thought, which has come to be known as the 'theory of action', is the object of widespread attention and debate in American sociology. Parsons has been a prolific writer, but the most concise summarization of his approach may be found in his *The Social System* (Chicago, Free Press, 1951). His very interesting *Essays in Sociological Theory* are available in a paperback edition (Chicago, Free Press, 1964).

An excellent introduction to reference-group theory and its broader sociological implications will be found in the work of Merton referred to before. The closest approximation to the understanding of reference-group theory as a nexus between role theory and the sociology of knowledge suggested by the present writer can be found in a direction-giving article by T. Shibutani, 'Reference Groups as Perspectives', published in the *American Journal of Sociology* in 1955.

CONCERNING CHAPTER 6

One of the best introductions to the methodological problems of the sociological enterprise is Felix Kaufmann, *Methodology of the Social Sciences* (New York, Oxford University Press, 1944), which can be supplemented by reference to John Rex, *Key Problems of Sociological Theory* (London, Routledge & Kegan Paul, 1961). Parsons's discussion of the relationship between Weberian and Durkheimian sociology can be found in his already cited *The Structure of Social Action*.

Part of Weber's writing on charisma can be found in Hans H. Gerth and C. Wright Mills (trs. and eds.), *From Max Weber* (London, Routledge & Kegan Paul, 1948). Carl

Mayer teaches sociology at the New School for Social Research. His work on religious sects, published in Germany in the early 1930s, is not available in English. An acute and detailed analysis of the origins, ideology and structure of three religious sects can be found in B. R. Wilson, *Sects and Society* (London, Heinemann, 1961).

Goffman's interpretation of the fate of being an 'inmate' is found in his previously mentioned work *Asylums*. His concept of 'role distance' has been further developed in his *Encounters* (Indianapolis, Bobbs-Merrill, 1961).

Georg Simmel (1858-1918) was another classic German sociologist. His approach to sociology is characterized by strong philosophical interests coupled with catholic scope in his analyses of various sociological problems. Simmel has been regarded as the founder of the so-called formalistic approach in sociology, continued in Germany after his death by Leopold von Wiese and others. The best anthology of Simmel's writings in English is Kurt H. Wolff (ed.), *The Sociology of Georg Simmel* (paperback edition, Chicago, Free Press, 1964). Lewis A. Coser's interesting discussion of *The Functions of Social Conflict* (London, Routledge & Kegan Paul, 1956) is based on principles drawn from Simmel's writings on conflict.

Johan Huizinga's provocative work, *Homo Ludens*, was published by Routledge & Kegan Paul, London, 1949. Maurice Natanson, a former student of Alfred Schuetz, now teaches philosophy at the University of California, Santa Cruz. A collection of articles by him has recently been published – *Literature, Philosophy and the Social Sciences* (The Hague, Nijhoff, 1962).

CONCERNING CHAPTER 7

The book by the present writer referred to in the text is *The Precarious Vision* (Garden City, N.Y., Doubleday and

Co., 1961). In addition to discussing the implications of sociological thought from the viewpoint of Christian faith, the book also develops further some of the ethical problems touched upon in this digression, especially as these problems relate to the sociology of religion.

CONCERNING CHAPTER 8

For important discussions of the role of sociology as a scientific discipline in the modern world, the reader may turn to Robert S. Lynd, *Knowledge for What?* (Princeton, Princeton University Press, 1939) and to Florian Znaniecki, *The Social Role of the Man of Knowledge* (New York, Columbia University Press, 1940). A recent statement on the same subject that is quite close to the viewpoint here presented (though it would probably not go as far as the present writer in understanding sociology as itself belonging to the humanities) may be found in Edward A. Shils, 'The Calling of Sociology', in Talcott Parsons *et alii* (eds.), *Theories of Society* (New York, The Free Press of Glencoe, 1961).

INDEX

MORE ABOUT PENGUINS
AND PELICANS

Penguinews, which appears every month, contains details of all the new books issued by Penguins as they are published. From time to time it is supplemented by *Penguins in Print* – a complete list of all our available titles. (There are well over three thousand of these.)

A specimen copy of *Penguinews* will be sent to you free on request, and you can become a subscriber for the price of the postage – 30p for a year's issues (including the complete lists), if you live in the United Kingdom, or 60p if you live elsewhere. Just write to Dept EP, Penguin Books Ltd, Harmondsworth, Middlesex, enclosing a cheque or postal order, and your name will be added to the mailing list.

Some other books published by Penguins are described on the following pages.

Note: *Penguinews* and *Penguins in Print*
are not available in the U.S.A. or Canada

THE FOUNDING FATHERS OF
SOCIAL SCIENCE

Edited by Timothy Raison

'A science which hesitates to forget its founders is lost'
– Alfred North Whitehead

The founding fathers of sociology continue to be widely read and revered because they have so much to tell us about sociology and ourselves. These essays on the men whose human inquiries have slowly established a new discipline with a scientific procedure and an intellectual independence of its own emphasize the importance of their pioneering work.

A RUMOUR OF ANGELS

MODERN SOCIETY AND THE REDISCOVERY
OF THE SUPERNATURAL

Peter L. Berger

An extraordinary essay by the American sociologist, who is conscious that his previous book, *The Social Reality of Religion*, might be taken for 'a treatise on atheism'. Here, however, poacher turns gamekeeper and Professor Berger suggests that fresh evidence of the supernatural may be found in things taken for granted in our daily life – in our faith in order, in play, hope, humour and even in our disposition to assign the worst evil-doers to 'damnation'.

'This little book, vigorous and authoritative, could mark an important turning-point in twentieth-century thought' – *New Society*